BECAUSE EACH LIFE IS PRECIOUS

BECAUSE EACH LIFE IS PRECIOUS

Why an Iraqi Man Risked Everything for Private Jessica Lynch

MOHAMMED ODEH AL-REHAIEF

WITH JEFF COPLON

HarperCollins*Publishers*

HarperCollins books may be purchased for educational, business, or sales promotional use. For information, please write: Special Markets Department, HarperCollins Publishers Inc., 10 East 53rd Street, New York, NY 10022.

FIRST EDITION

Printed on acid-free paper

Library of Congress Cataloging-in-Publication Data is available upon request.

ISBN 0-06-059054-8

03 04 05 06 07 ❖/RRD 10 9 8 7 6 5 4 3 2 1

For my parents,
and for the United States Armed Forces

ACKNOWLEDGMENTS

B OB LIVINGSTON has gone above and beyond the call to help me make a new life in the country I now call home; I owe him more than I can tell. In The Livingston Group I found a second family, and I especially want to thank J. Allen Martin, Jim Pruitt, Dick Rodgers, and Lauri Fitz-Pegado.

Rita Kantarowski and Nada Ashkouri of the International Rescue Committee did so much to ease my family's transition into the United States. Marvin Josephson, an IRC board member and founder of International Creative Management, gave selflessly of his time and resources to make this book possible. My agents, Esther Newberg and Carrie Stein, freely offered their wisdom and encouragement, and were always there to answer my questions. David Hirshey, my editor at HarperCollins, did wonders in helping to shape and improve every page of the manuscript; his assistant, Nick Trautwein, made sure that we met our difficult deadlines.

I must also thank George Hishmeh, my translator; the staff at the Berlitz Language Center in Washington, D.C.;

Sean Murphy, consul at the U.S. Embassy in Kuwait; Dr. Mark Johnson, my retinal surgeon; Greg Reid, Tony Flores, and Pete Hammond, our bodyguards in our journey from the Middle East to the United States; Lt. Col. William Perez, who gave us indispensable help in clarifying the military's side of this story; Col. Michael M. Wyka, USAF, who took such good care of my father and the rest of my family; Senator Joseph Biden, who lent his assistance to get my family out of Iraq; friends Andy and Nancy Ireland, Hal and Donna Cohen, Adnan A. Saad, William and Sharon Murphy, Melissa Sara, and Paul Gaubatz; Tom and Mary Khalil; and the Marines, Sailors, and Soldiers of Task Force TARAWA (2nd Marine Expeditionary Brigade), who made *everything* possible.

Finally, I must thank Lamiyae Jbari of The Livingston Group. She has been my trusted colleague, my voice of reason, a wonderful translator, and a great friend. She has supported this project every step of the way; I truly could not have done it without her.

Mohammed Odeh al-Rehaief
AUGUST 26, 2003

BECAUSE EACH LIFE IS PRECIOUS

PROLOGUE

GROWING UP, I was known as "a long tongue." (Loosely translated, you might say "a big mouth," or maybe "a wise guy.") I had an opinion about everything and could not keep it to myself. Plus I was willful—I stepped without thinking, leaped before I looked. In the age of Saddam, these were bad habits for Iraqis, even young ones. I made my poor baba gray before his time.

"It's not normal," my father would say, shaking his massive head, raising his dense eyebrows. "There is always some trouble with you, something unexpected. Your ambitions are too big for your size. If you don't stop looking up so high, you are going to break your neck."

When I was six years old, at the start of my very first day in school, the teacher called us to stand and face the blackboard. Next to it hung two portraits of military men with all the trimmings: epaulets, gold braid, silly hats. One was a kindly-looking, gray-haired man: President Ahmed Hassan al-Bakr. The second was a younger, fleshier fellow, his right hand raised in a tight salute. Though still vice president in 1975, Saddam Hussein was the most powerful force in Iraq, and his picture was everywhere. Led by our teacher, my tiny classmates clapped in time and chanted: *"Welcome, my father Saddam! Compassion grows in you day by day! We are happier when we see you!"*

I thought they were all lunatics—what was *that* about? After they finished, my *austada* loomed over me: "Why aren't you clapping? Did your parents tell you not to clap?" (These questions were more sinister than I knew. Some teachers tricked students into informing on their families, with tragic results.)

"No, no," I said. "I just didn't understand why we should clap at a picture."

The teacher frowned; she loved Saddam. "You need to clap," she said. "This time I will not punish you, but next time you'll get a spanking." Soon enough I'd learn what that meant: a wooden ruler across my tender palms.

At home that night, when I told my parents of this puzzlement, Baba was dismayed. He nipped my rebellion in the bud: "You must *always* clap, or they will come and lock us all in jail. And in jail, you do not get good food."

My father knew how to scare me, all right. No *masgouf,* my favorite oven-roasted river fish? No *mtabak dajaj,* my beloved chicken and rice? No *ketchup*? There was invariably a bottle of ketchup at our kitchen table, as I would eat nothing without it. I imagined a first-grader's hell: a life without condiments.

The next day I clapped with the rest. But though I mouthed the words, I did not chant. For once, that long tongue of mine was still.

PART ONE

CHAPTER ONE

BY MID-MARCH, the winter chill and damp gives way to spring in our part of the world. The season came on time this year, but it wasn't quite as usual. The sun shone, the lilies bloomed . . . and American missiles took aim at the presidential palace in Baghdad.

My hometown of Nasiriya was a market center of 300,000 people, about 200 miles southeast of the capital. Most of us were Shiites, Iraq's left-out majority, the ones who stewed in hidden hatred of Saddam Hussein. Like my neighbors, I had seen too much war in my thirty-three years. We all knew its price; we knew that innocents would suffer. Even so, most of us were eager for the U.S. to rid us of Saddam and his Baath Party. Our fear was that the Americans would do the job halfway—that they'd stop before Baghdad, as in the Gulf War.

On the night of March 17, when George W. Bush gave Saddam forty-eight hours to leave the country, some Iraqis thought their leader would run. I never believed it. Saddam was too crazy and pigheaded to back down. He dreamed he was a Saladin who would turn back the infidels and protect our ancient civilization. He did not see himself as a thief and a gangster, a thug in president's clothing. No one near him dared hold up a mirror.

With each murder—and there were untold thousands of them—Saddam had fathered more enemies. Without his Republican Guard, his legions of security and intelligence and paramilitary goons, there was no place in the world he could be safe. He would cling to Iraq to the end. Our one hope to avoid war, I thought, was for the Americans to target a building with Saddam inside. A *decapitation,* as the CIA director called it.

But when a snake slithers underground, it is not so easy to cut off its head.

The day after Bush's ultimatum, I joined my two younger brothers, Ahmed and Hassan, in the lines of hoarders at the souk, the community market. Our family farm, twenty miles to the south, would keep us in dates and molasses, but the rice harvest was sold and gone. We would also need flour and sugar, and chicken and lamb to stock the freezer at Baba's house in town, where my brothers lived with their families.

The first air-raid sirens came on Thursday, March 20, at around 10 A.M., when I had just left home to take my car for an oil change. I raced back and led my wife, Iman, and our five-year-old daughter, Abir, to the airtight basement beneath our home. Every family of means in our town had one. If Saddam was pushed to the wall, we assumed that he would wage chemical warfare and the United States might retaliate in kind.

Our shelter had a generator, an air vent and charcoal filter, a refrigerator and stove, food and water for twenty days. There was a bed for each of us and a spare one for a guest. Every seam was lined with a sealant used by goldsmiths.

Iman fixed us tea and we sat for two hours. We had a

sturdy stone house, but the bombing rattled it to the foundation. Hungry for news, I tuned our radio past the official Iraqi stations; our information minister, Mohammed Said al-Sahhaf, was an utter clown. I stopped at Radio Sawa, an Arabic-language station funded by the U.S. government. I heard about fierce fighting in the far south near Basra, where my brother Ali worked as a dentist.

In those first long hours of the war, as we waited out the siege, I feared the unknown dangers ahead. But I also hoped against hope that our time had come at last. I hoped that we might win a future for our children, free of a regime that tracked their every move and word.

Abir refused to leave the shelter when the all clear sounded, so we brought down her toys. Iman would stay with her, skipping her nurse's shift at Saddam Hospital at the northwest edge of the city. I called my father, who asked me to check the farmhouse and see if it might be better for all of us there. Half an hour later, I was hugging my sister Fatma, who lived with her husband on the farm.

"It is not safe here," she told me. "There is army all around." Saddam's troops were camped among the date palms. It seemed unwise to move next to them.

Back in town, Baba called a meeting. "Children," he told us, "the war has started." To protect the family businesses, he issued his marching orders: Ahmed to the hotel, Hassan to watch the clothing and perfume stores downtown. I would stay with Baba. In the meantime, we would send our wives and children, including Ahmed's two young daughters, to our vacant rental house. It was next to a mosque and not much else. There they might be safe, Baba thought.

I had a different idea. With no furnishings, the rental house seemed impractical. And judging from what happened in the Gulf War, I needed to keep an eye on my own place to fend off looters. Anyway, who could say where safety resided in a whirlwind? There is an Arab expression: *What is going to happen, will happen.*

I drove back across the Euphrates and down Baghdad Street to our home in southwest Nasiriya. We settled downstairs for the night. Deaf to the cruise missiles and guided bombs, I fell into a deep sleep—until Iman jostled me awake, saying, "Talk to your daughter." Abir was trembling with fear. When a bomb rocked our house, she would cry, "Oh, I'm going to die!"

Trying to comfort her, I said, "It's not an air raid. It's only the thunder."

She looked at me skeptically: "It's not raining." A moment later, a near miss knocked a pot off our stove. "*This* is thunder?" she said.

The next day was Friday, the Muslim Sabbath and the start of the ground war. People crowded the mosques in hope of warding off misfortune. But the barrage continued around the clock. On television they announced Saddam's bounties: 100 million dinars for shooting down an American or British plane; 75 million for downing a helicopter; 50 million for killing a Marine; 25 million for capturing a Marine; 15 million for the capture of any Iraqi found helping the occupiers. Even with our devalued currency, at about 2,000 dinars per dollar, these were powerful temptations.

On Saturday, March 22, a stray bomb obliterated a house just fifty yards from Baba's. The owner, a businessman, had fled the city with his family. But a carpenter's eight-year-old son, named Malak, was not so lucky. He was playing by the stone wall outside the house when the bomb arrived.

An intriguing boy, this Malak. He could neither hear nor speak, yet he was very bright. His father told a story about a pendulum wall clock that chimed on the hour until it broke. The father took the clock apart and tried to fix it, to no avail—it would have to go to the shop. He left the pieces on a table and went off for a nap.

An hour later, the father awoke to a telltale chime. At first he thought he was dreaming. He went back to the living room—and there he found his son, transfixed by the rocking pendulum. The clock was on the wall, keeping perfect time.

The carpenter's wife said, "Malak fixed it." She had seen her son with the parts and thought he was just playing. She wanted her husband to get a new clock, anyway, so she'd left Malak alone to fiddle.

Who knew that the boy was a genius?

Did Malak stay outside that day because he didn't hear the siren? Did someone call out a futile warning? At the end, there were no explanations that mattered. Neighbors shielded the parents from seeing the remains, which Hassan took to the morgue.

When I heard what had happened, I felt that we were headed for catastrophe—that someone dear to me would die. My wife, my daughter . . . or perhaps I would meet my own end. If the war could claim a sweet boy like Malak, who among us owned the right to survive?

The crowd on Baba's block buzzed with rumor that day. One neighbor insisted that the bomb was not American. He had seen a fragment with some Russian letters on it. He suggested that the Baath Party was responsible.

That night, I was told, the Fedayeen Saddam—the president's "Men of Sacrifice," his loyal paramilitary, the most ruthless, vicious people in Iraq—came and took the indiscreet neighbor away.

wow.

The Americans were not the first invaders to meet stiff resistance in Nasiriya. The city squats on the Euphrates at a strategic point, a third of the way between Kuwait and Baghdad. Back in 1915, when the British took on the

Ottoman Empire there, five hundred men were lost on each side. Now the coalition was retracing those steps along Highway 8, the straightest line to Baghdad. Its supply lines to the front crossed the Euphrates and the nearby Saddam Canal over two bridges at Nasiriya, a few miles east of my home. For the Americans to capture the capital in good time, those lines had to be secured.

On the other side, the Iraqi commander for the southern zone, Ali Hassan al-Majid—better known as "Chemical Ali"—decided to make a stand in our town. Snipers were posted on every other rooftop, and the fedayeen were five hundred strong or more here. A pitched battle would flare for more than a week. No part of Iraq would be more treacherous than the stretch between those two bridges. It became known to the Americans as Ambush Alley.

In Arabic, *nasiriya* means "victory." The city's people are clannish and wary of outsiders. They have put up with centuries of injustice—from the Turks, the British, and finally the Baathists, a clique that favored Sunni Muslims over Shiites at every turn. The Shiite revolt of 1991 broke out in our city first before spreading throughout the south. Saddam took a terrible revenge here, killing and torturing into the thousands.

Nasiriya had its share of fundamentalists. Some could be inflamed by the regime's call for jihad against Bush and Tony Blair. But you must also know that Ali Hassan al-Majid formed death squads to hunt down deserters—to hang them or, if time was short, simply shoot them in the back. Many Iraqi soldiers were caught in the middle. They had no love for Saddam, yet no place to go.

For the United States, the bloodiest day came on Sunday, March 23, and most of the blood was spilled in Nasiriya. Nineteen U.S. Marines were killed here that day, and fifteen more wounded. Early that morning, the tail end of a vast U.S. Army supply convoy failed to get word that it had been rerouted to the west, to skirt the city and its militias. Several wrong turns later, this band of clerks and cooks was set upon just south of the Euphrates, on the outskirts of the city. The Americans came under fire from all sides. Their rifles and machine guns jammed. A five-ton tractor trailer stalled in soft sand by the road. A Humvee, rocked by a grenade, swerved smack into the trailer. It was a lethal crash. Only one of the Humvee's five occupants would live to see the next day.

In that ambush, in that moment, many lives would change forever.

Including mine.

In between air raids, I left the house to trade gossip at the street corner with my neighbors. We guessed at how far the coalition had driven, and what Saddam would do next.

As a stream of people surged past us, I stopped a man to ask what was going on. Breathless, he told me, "They are bringing American prisoners to the Baath Party office!" These were the convoy survivors, the ones who had been spared.

The local Baath headquarters was less than half a mile from my house. When I arrived, I found five hundred people overflowing the street and sidewalk, stirring with excitement. They were about to see something new. Party officials swaggered by the building's door, looking important. The air hung heavy with anticipation.

The show began half an hour later. Two late-model white pickup trucks, flanked by four blue Land Cruisers, pulled up to the headquarters. I could see two American soldiers in the bed of one of the pickups, three in the other. Except for one man in a stretcher, the prisoners were seated with their hands tied behind their backs. Their heads were bent, their eyes blindfolded.

The viewers, so raucous a minute before, fell still. A

thousand eyes stared as the Americans were led from the trucks. Their blindfolds were removed, and two were forced to haul the stretcher. Their faces looked bruised, as expected—that was the practice for prisoners in Iraq, foreign or domestic. As the security men steered them toward the building, I saw that one soldier was a black woman. She must have been injured; she walked with a limp.

Then the storm broke. A middle-aged Iraqi woman, after pushing her way to the front row, flailed at the Americans as they passed, screeching oaths. A younger man, apparently her son, flew in from behind and screamed, "This is for my brother!" as he struck at them. From what I could tell, the brother had been slain in the early fighting.

The guards quickened their pace. The woman fell back, but the son kept pounding the prisoners' backs until they got inside the door. There they would be processed and photographed.

Forty-five minutes later, the door opened and the prisoners filed out. Now the guards did the pounding, arousing the crowd. The Americans were courageous. When beaten, they made no sound. As the trucks roared off to some unknown place, they were at the hands of men who knew no rules. Yet they seemed stolid, even confident. I could not help but admire these soldiers. Would I do as well in their shoes?

A few hours later, toward sunset, I climbed to my roof to see what had been hit. A loudspeaker crackled through the air: "Here are the dirty bodies of the Americans!" Then I saw the parade, slowly rolling my way toward the roundabout on Baghdad Street, a block from my home: seven Toyota pickups in a neat file.

Telling Iman and Abir to stay put, I hurried to the roundabout. By the time I got there, the three lead vehicles—holding military police and fedayeen—had reached the straightaway beyond me. The next truck was passing, and all I saw of those corpses were three or four pairs of bare feet.

Truck number five I saw squarely—it was hard to look, yet harder to look away. The bodies of two black men were sprawled in the back. Between them slumped a young brownskinned woman, her dusty torso at a forty-five-degree angle to the floor. Her legs were splayed apart, her pants rolled up to her knees. All three bodies were in soiled, brown flannel tops. Their uniform shirts were gone; the looters had been busy.

Many days later, I would learn that the young woman was Private First Class Lori Piestewa, the driver of the ill-fated Humvee.

A black-clad fedayeen struck a pose of arrogant triumph. He stood on the chest of one dead man and the head of the other, straddling the body of the woman. I thought of Abir, and how this soldier had been a little girl not so long ago.

Standing next to me, two older women in *abayas*, the traditional long overcoats, clucked at this display. They were most offended by the female soldier's bare legs: "What kind of Islam is this, where you show the woman's body to everyone?"

The fedayeen blasted their Kalashnikov assault rifles into the air. Eighty or so people ringed the roundabout, most of them cheering, "God keep President Saddam! God give President Saddam long life!" (When Saddam was in power, it was not permitted to use his name in public without the title of "President" or "His Excellency." If you were sloppy about it, you could get six months in jail.)

In a Shiite town like Nasiriya, not too many people loved the regime. But the citizens of Iraq were well trained. They saw the two trucks at the rear of the parade: a white one with Baathists and a gray one with press photographers, snapping pictures en route. Everyone knew it could be fatal to lack enthusiasm at such times.

Past the roundabout, the pickups cruised down Baghdad Street, toward Victory Bridge and Saddam Hospital. From Hassan's wife, Hamida, a doctor who worked there, I would learn that the Americans were lugged roughly from the pickups by their legs, or heaved tumbling to the ground. A hospital security official kicked several bodies in the head. Two fedayeen brawled over a soldier's body armor, like jackals over carrion. Their superior stopped the fight by grabbing the armor for himself.

As doctors and nurses watched with revulsion, the bodies were finally deposited in the hospital morgue, a place already filled to overflowing.

Monday morning, thanks to generator power, we found Saddam on state television. "Victory is near!" he proclaimed. "Strike against your enemy without fear, you great Iraqis, inspired by the spirit of jihad!"

I laughed out loud. Saddam had lost all touch with reality. I knew from Radio Sawa that the Americans and British had taken Basra. They were advancing on Karbala and Najaf, and there were battery movements in the north. That made four fronts of attack against the regime, with far superior technology. It seemed only a question of time before the state collapsed.

Minutes later, they replayed a news clip of the American POWs. It was ugly to see them sitting there, taunted by their captors. It was worse to see the corpses on crude display. I worried about what the world would think of us. Would the American public turn against the war and force Bush to withdraw? Would the U.S. military treat common Iraqis more harshly?

As I watched the faces of the POWs, they seemed younger and more bewildered than I remembered. The black woman looked especially terrified, and I could not blame her. I thought of how each of them had left mothers, spouses, children back home. Would they ever make it back?

Two days earlier, the information minister said that any American who "landed in our hands" would be treated as an occupier and eliminated. But Saddam announced a change in policy. Now the POWs would be interrogated and saved for prisoner exchange.

You never knew with Saddam, though. After the Gulf War, he declared an amnesty for deserters who turned themselves in—and then "amended" the offer just in time to torture or kill those foolish enough to take his word. As long as the American prisoners were under his control, I knew that their lives were in danger.

The bombardment of Nasiriya came yet heavier that day, with artillery to root out resistance and break the local stalemate. Rather than venture into the war zone, my wife stayed home from her job a second day. Someone noticed. That evening, an ambulance driver stopped by our house to ask Iman, "Why didn't you show up at work today? Ali Hassan al-Majid was there, and all the government bigwigs." Iman tried to explain, but the driver cut her off: "You better come in tomorrow." It sounded like a threat, most likely from the hospital director himself.

In wartime, people who skipped work were shunned as cowards, almost deserters—and at Saddam Hospital, as Iman had seen firsthand, they cut off deserters' ears. Chemical Ali was the president's cousin, the hatchet man's hatchet man. In 1988, he had boasted of liquidating 100,000 Kurds with nerve gas. Three years later, he did Saddam's dirty work in the south, squashing the Shiite rebellion with mass arrests and torture. He was not known for his patience or compassion.

The next morning, my wife and I agreed, I would drive her in to work.

CHAPTER TWO

T HE FEDAYEEN had suffered not a single casualty in
their attack on the convoy. Emboldened, they beefed
up their resistance along Ambush Alley. The U.S. command
was losing patience. On Tuesday, March 25, a column of
4,000 Marines bulled their way through. Through a hail of
mortar and rocket fire, battling a sandstorm, they crossed
the two key bridges over the Euphrates.

To the west, Victory Bridge remained in the regime's
hands. Knowing its importance, the Americans were shelling
it with regularity—not to destroy it, as they had during the
Gulf War, but to discourage Iraqi troop movement.

Iman's commute was usually packed at rush hour, but
we seemed to be the only civilians out that morning.
When we reached the bridge, just before eight o'clock, the
pavement was cracked and cratered from the shells. Only
one lane was passable. Vehicles on either side lined up to
take turns crossing.

I inched my Volvo behind a jeep; a canopied truck, with
nine or ten fedayeen visible in the back; and a Baath Party
pickup. As we waited our turn, we saw a white Land

Cruiser glide past us from the opposite side, then a second one. The hearsay was that these carried Chemical Ali's death squads. They were ghost cars, best avoided.

On the other side, as we drove through downtown, the damage was worse. Power poles were down, their dead wires webbing the asphalt. Smoldering wrecks of cars dotted the streets, which were eerily empty. There were no sounds except the distant boom of large guns—and the concussion of the shells, much closer, as they landed.

A few blocks from city hall came an earsplitting blast. Bricks flew by our windshield. We looked up—and the top two floors of the Governate, our city hall, were no longer there. Maybe this was a bad idea, I said. Maybe we should turn back.

Iman would not hear of it. She'd felt bad about staying home even before the driver came around. It was her duty to go and treat the wounded. She said, "Put yourself in the place of somebody whose father was in that building, and now he has been hurt and goes to the hospital, and there is no one there to help him. If I didn't go in, my friends didn't go—how would you feel?"

Iman was a deeply feeling person, a born healer. (I, on the other hand, had been trained as an attorney; I worked from my head, not my heart.) So we made a compromise. We'd find a safer place to wait out the shelling, then proceed. I knew just the spot, the parking garage by our favorite movie theater, where we'd seen *Rocky* and *Batman* while Abir slept between us.

But the Iraqi army had claimed the theater for a garrison, and the garage was shut tight: military only. We pressed on and somehow made it to the hospital. I kissed my wife good-bye. It was hard to let her go.

On the way to Baba's house and my waiting daughter, I passed the local security center and the military emergency building, both gutted and charred. Adjacent homes and stores had been hit, with some still licked by flames. *There are no clean wars,* I thought. Would this one be worth it?

On one desolate stretch, I was startled by a woman who dashed into the street to stop me. She wore only her long yellow *jalabia.* Her black hair was free and wild, her feet bare. She was battering her face in grief.

"You must help me," she pleaded. "They have hit our house, and my family is still inside!"

I asked if she was injured—should I take her to the hospital?

"No," she said, as if I were missing the obvious. *"They* are injured. Come and get *them* to the hospital."

She got into my car, pointed me around the corner, ordered me to stop. She jumped out and ran ahead. I followed her to the house . . . but there was no house. There was only a mound of stone, a haze of dust, and a gaggle of onlookers. It seemed impossible that anyone could be alive in there.

Still, I had to do something for this poor woman. I went to one of the neighbors and asked him to call an ambulance. He looked at me quizzically and said, "Who should we call the ambulance for? Everyone is dead."

I tried to shut him up: "How can you say that?"

But the woman had heard. She began laughing hysterically. *"So, Abu Amar,"* she chided the neighbor, "they're all dead now?" By the tone of her voice, I knew she had lost many children. Abu Amar fell quiet. The woman laughed and laughed. Giddy, she picked up a clump of earth from her garden: all that was left.

Feeling a stab of worry about my own child, I drove on. My father's neighborhood was sandwiched between the two armies and getting raked from both sides. A half mile from Baba's house, I ran into a roadblock and a wall of choking black smoke. A gas station had taken a direct hit. With the fire brigades out of action, the blaze would have to burn itself out.

I spent the day with my family, shuttling in and out of their shelter. With each all clear, four-year-old Doha, Ahmed's younger daughter, would scamper outside to scavenge for shrapnel. She'd tote the jagged metal back into the shelter, delighted with her find. As the shards had dropped from the sky, she figured she could piece them together and build a plane. She could fly us all out of the war.

Abir found no such distraction. She would cringe at each explosion and say, "Oh, this is falling on my mother!" To calm her, and keep my own nerves at bay, I called the hospital every hour. Iman was glad she had gone in. More than half the staff was missing, and the rest were flooded with casualties. She'd ask me, "Do you remember so-and-so? Well, they just brought him in—he is dead." It happened several times that day.

At five o'clock, Hassan went to pick up our wives. Iman came back despondent at the sight of so many young lives ruined. She told us of some "prominent person" being treated in the cardiac unit on the second floor. It was likely a high Baath official, she thought. They'd placed guards outside the unit's entrance, and no other patients were being admitted there.

So typical, I thought. The whole country was bleeding, and here beds were left empty to please the mighty.

We stayed at Baba's for the night, to be nearer to the hospital. We left the generator idle and ate dinner in the dark, as any light might be targeted. Ahmed's wife, Wafaa, made *mtabak dajaj* on the gas stove. It looked delicious, but the chicken didn't taste quite right—it must have spoiled. Our three girls, too drained to eat, had crawled under blankets to sleep.

We bundled them down to the shelter, where mattresses crowded every inch of the concrete floor. The women and children slept farthest inside. Baba lay in the middle, while my brothers and I settled by the outer wall, where the risk was greater should a bomb fall close. Someone had to sleep next to the door, and my brothers voted for me: "You are the oldest one here, after all."

I might have been content if Ahmed had not dreamed of romance that night. He threw an arm around me, then a leg . . . and there I drew the line. I cuffed him awake and snapped, "Go lie with your wife!" He switched places with Hassan, who snored and kicked me in the stomach.

I gave up on rest even before the artillery resumed. When the noise awoke the children, I fell back on my thunder story. But Abir was still not fooled. "This is not thunder," she'd insist. "This is shelling!" That settled, she and her cousins began to wail.

The battle for Nasiriya raged on. On Wednesday, March 26, as Iman and I set off for Saddam Hospital, we saw a landscape of smoke and devastation. The few people in the streets were packing cars to leave the city. Should we be doing the same?

Halfway to our destination, we came across two men lying in the middle of a broad avenue called Sharea al-Guitarra: Guitar Street. One had been ripped at his middle, spilling his insides. The other, still alive, was facedown and thrashing his legs in agony. We jumped out of the car, thinking we would bring him to the hospital. But when we turned the man over, we saw the deep wounds in his neck and chest, the arterial blood. He stopped breathing in our arms.

Watching him expire unnerved me. I told Iman, "I don't care what they threaten—you are not going to work today. If you or I wind up like this, who is going to help us—Saddam? Let's survive today and think about your job tomorrow."

My wife saw it would be no use to argue. We turned back and collected Abir, then drove to our house as fast as the streets allowed. At Victory Bridge we found a single jeep waiting in front of us. Over the water, a pair of looters shuffled in our direction on the walkway, hauling a pushcart with a TV and a small refrigerator. As one pushed from behind, the other tugged from the front and steadied their spoils.

They made slow progress—too slow. The bomb killed one of the looters on the spot, hurling him into the roadway. His partner, in a stained white *jalabia*, astoundingly got to his feet and crossed over the bridge, hobbling past us without a glance.

By reflex, I'd covered Abir's eyes when the men were hit—but had she seen it before I could react? She burrowed her head into her mother's lap, rolling her body into a ball. Abir was a slight child, only fifty pounds, but of late she could not get small enough. She had been at the edge for too many days. Her mischievous smile was a dim memory.

It was good that my daughter did not see what happened next: the jeep running over the dead man's legs. It sideswiped the cart, bouncing the appliances over the rail and into the river. When our turn came to cross, I carefully veered around the body.

I was ready to take to my bed, but a new problem needed fixing. Shrapnel had pierced the tank on our roof, leaking away our water. I carried a picnic cooler to the end of the block, where people in similar straits were filling all sorts of vessels from a neighbor's hose. We took for granted that whoever had such necessities would share them.

It was a long line, and frustrating duty; I could get enough water for bathing or cooking, but not both. I watched an ambulance idling at the curb for no apparent purpose. Why was it here? Weren't there enough emergencies in Nasiriya?

I asked the next man to save my spot, and went to get Iman to relieve me. At the gate to the next house, I saw a fedayee named Abbas at his door. I was about to say "Salaam Aleikem" when he turned to speak to his mother inside. I was curious about Abbas, who had driven one of the pickups holding the American POWs. I knelt by his gate to tie my shoes and eavesdrop.

The mother said, "Will you come back for dinner tonight?"

Abbas said, "No, I told you—we're doing the drive-by."

The mother grew agitated. "You'll get killed," she said. "Let the others do it. Don't go."

"I'll be fine. And don't worry about dinner—I'll be out with my friends."

"Do you have any money? I need six thousand dinars."

After Abbas counted out the bills and turned down his walk, I straightened to shake his hand. I told him, "You men are heroes! Where are you going tonight?"

Abbas was a broad-shouldered young man with a large head and short, curly hair. He was not stupid, exactly, but he was vain and a show-off. He loved running his mouth, especially when trying to impress his girlfriend, Huda, who stood preening by her house across the street. He pointed to the ambulance and said, loud enough for Huda to hear, "We have a suicide bomber inside—we're going out to kill the Marines."

I could see the men in doctors' gowns inside, waiting for their friend. "You're a lion-heart!" I said. "Where will you do it?"

Abbas told me they were bound for the Mansuriya district, to the east. "Yeah, it's dangerous," he said, "but if I don't do it, who will defend our country?"

"What time are you going there?"

Abbas arched his brow and said, "Why are you asking? Why do you want to know so much?" We'd been neighbors for years, but he would cheerfully run me through if he thought I had it coming. Still, I wanted to get more out of him. In these uncertain days, information was gold. You never knew when it might buy some time, or your life.

Smiling, I said, "I just wanted to tell people about the great things you are doing."

The fedayee thought a moment, sizing me up. He said, "I don't know for sure—either tomorrow morning or the next day. We'll wait until dawn, when those dogs are sleeping."

We shook hands again before he loped across the street, to tell Huda not to bother standing in line for water: "I'll bring a gallon to you later!"

The shelling worsened toward sunset. After hearing a strange new sound, a deep vibration, I went back outside to a chilling scene. Companies of Iraqi soldiers had fanned out across the neighborhood. Tanks rumbled through the streets. Our block was a defense line against the Americans! We needed to get out, immediately. Iman packed her jewelry and a few other small valuables, and we took Abir to the car.

We did not get far. They had blocked off the bottom of our street, and a guard sneered at my story that we lived across the bridge. "No, you are Mohammed, the lawyer," he said. "You live right over there. You have to go back."

In a city like Nasiriya, in a police state like Iraq, no one was anonymous.

I could feel Iman panicking. As I closed the garage door

from the inside, I said, "Don't worry. We're fine. Nothing's going to hap—"

I had no way of knowing that some unknown Iraqi had climbed to the roof of a vacant house, three doors from our own, and fired a rocket-propelled grenade at a U.S. helicopter. The Americans answered with a missile that found its mark. I *felt* it first, a quake that nearly threw me off my feet. The noise was monstrous.

Iman clutched my shirt and yelled, "Let's go!" Her voice came from deep in her throat; her eyes were black and huge. Abir was paralyzed. I took my daughter's hand, but she refused to come inside the house. I tried to explain that it was too dangerous outside. There would be shrapnel and cross fire. We could not risk it.

"We're going to die!" Iman screamed, pushing me in the chest. "We're going to *die!*" It was a tense moment. We couldn't stay in the garage; the aluminum door would make a flimsy barrier against some projectile. Knowing I had to take control, I twisted Iman's wrist and pulled her to my body. I forced her inside to the family room, with a sobbing Abir behind us. I could see into the kitchen, where the windows had blown out. I kept pushing my wife down the stairs to the shelter, where I doused her with a pitcher of water and tapped her on the cheeks.

Iman said, "What happened?" She had no memory of the last minutes—including my rough treatment, thanks to God. The streets were still again; the soldiers must have pushed on. We moved back upstairs.

That was the evening of refugees in our neighborhood. They came from the city's outskirts to the east, a place called Zenawiya, where the American offensive and Iraqi counterattacks had driven them from their homes. When three bedraggled people showed up at our door, we did no more or less than anyone else. We gave them food and drink, water to wash the mud from their bodies, shoes for their bare feet. We did what we could.

There was an older man, thin and dark, with a twisted mouth; a good-looking younger man; and a woman about forty, with a moon face. The older one, named Abjabar, told us their story: "When the Americans got to us they were firing, and I knew they would not be merciful because of what we had done."

I asked him, "What did you do?"

"Not *us*," he said. "The fedayeen and Baathists—the ones who'd killed the American prisoners." *Those must have been the dead soldiers,* I thought, *the ones paraded in the roundabout.* Early that morning, as Abjabar told it, the paramilitaries had come into their neighborhood and lined both sides of Highway 8, the road that led south from the city and Ambush Alley. Alarmed, some residents left their homes and hid in a grove of palm trees. A loudspeaker strapped to a truck ordered them back: "Get inside! Get inside your houses!" They did as they were told.

Abjabar went up to his roof to see what was happening. Before long he saw the approach of the convoy, the lost trucks and trailers of what the world would come to know as the 507th Maintenance Company. "We were surprised," he said, "because the Americans were not shooting. They had no tanks with them. Had they come to *watch* us? When they got to our area, the fedayeen surrounded them and shot at them."

The woman interrupted: "There was an accident between two of the trucks."

"There was an accident," Abjabar agreed. "The Baathists and fedayeen fired on the trucks as they approached. They shot the driver, a woman soldier. Another woman, she was fair-skinned, they did not shoot her. They took her from the backseat of the truck to the ground. They were stomping her with their rifles. Then they took her away.

"An injured soldier limped out of his truck. As soon as he straightened up, they shot him. Another American stayed inside his truck and waved his hand out the window to surrender. They took him, too."

The old man sipped his water and said, "I was unlucky. I wanted to go fight and kill the Americans to get the reward, but they did not let me."

I was unsure how much to believe from Abjabar—how much came from his own eyes, how much from gossip. He'd observed the fighting from a distance, and it must have been chaos. His bluster aside, I knew that he'd been afraid. Still, I suspected that the woman he'd seen shot was the one I had observed at the roundabout.

And what of the second woman, the fair-skinned soldier, the one Abjabar said was beaten? What had happened to her?

CHAPTER THREE

BY THE EARLY MORNING of Thursday, March 27, the coalition had encircled Nasiriya from the east, west, and south, but they still had much work to do. Shortly after midnight, a fedayeen company launched a rocket attack. They injured twenty Marines before they were beaten back.

As dawn approached, Iman and I fought sleep in our shelter. I thought about using my lawyerly skills to get a brief leave of absence for my wife from the hospital. We might move to the farm if the troops were gone. And if not? Solutions eluded me that night.

We wound up oversleeping—it was eight-fifteen, and Iman was already late. Mindful of the food shortages, we had only bread and butter and milk for breakfast. Except for Abir, of course, who had eggs and cheese and jam. But none of it helped when she saw her mother changing into her nurse's whites. She begged Iman to stay with us; I steeled myself not to beg along with her.

We left the house at eight forty-five and stopped dead two blocks later, out of gas. The warning light had flashed the day before, but the stations were closed and I'd hoped to make it to my father's before running out. As I began walking home to siphon some gas from Iman's Nissan, a woman ran past me, shrieking, "God is great! Will God accept this?" She was trailed by a dozen more people, all in consternation. I saw a neighbor, a barber named Hallaq, and asked what was going on.

"Are you sleeping?" he said. "Didn't you hear the shots? Hind was killed!"

"Why was she killed? What did she do?"

"It was the fedayeen."

I was shocked. Hind lived two blocks from us with her three boys—the youngest, I recalled, was only six. Divorced, she supported her family by selling railway tickets at the station on our side of the river. She was friendly and helpful. I could not think of anyone who wished her harm.

As Hallaq and I talked, the people sprinted the opposite way. Hard behind them, one of the ominous white pickups sped over the pavement. It dragged a lifeless woman, her face to the sky. They had tied Hind by her wrists to the rear fender. She was soaked in blood. Around her neck they'd tied a sign, scrawled with a black Magic Marker: THIS IS THE FATE OF A TRAITOR.

Of the four men in the pickup, I recognized only the driver, a young fedayee named Mehdi. He was Hind's next-door neighbor.

Struggling to catch his breath, Hind's father stopped and joined us, with his grown son alongside. The father was frail and elderly, an uneducated man. Hind was the joy of his life. When others came with their sympathies, he angrily waved them off. He said, "Nobody gives me condolences! I know how to take revenge!"

His son said, "Where are they going to run from us? We know who they are. We will kill them all!"

From others I pieced together what had happened. Everyone knew Mehdi as a hard, despicable man. He once had a brother who hated Saddam and planned to flee to Iran. Mehdi turned him in, and the brother was executed.

It turned out that Hind and Mehdi's mother had squabbled. In retaliation, the mother told Mehdi that she had seen Hind waving to a Marine Cobra helicopter—after setting a red blanket on her roof to point to Mehdi's house. The family knew that Hind gathered her water near a Marine base. They decided she had hatched a plan to target them.

When Mehdi went to his group of fedayeen from outside the neighborhood, he embroidered the tale. He said he had actually seen Hind going to the Americans, enough to sentence her to death. They came to her house, shot her where she stood, and brought her body to the pickup.

None of Mehdi's story was true, except for the fact that Hind had left her blanket on her roof—to air it out, in the Arab custom. The murder was an outrage even in this savage time. All Nasiriya heard about it. They put it on Radio Sawa.

Here were two tragedies, I thought. A blameless mother had been murdered, and Mehdi's house was not bombed. Was there no humanity in Iraq? No justice?

Could a man live here without losing his soul?

We dropped Abir at Baba's and got Iman to her job at ten-thirty. Saddam Hospital was a six-story, white-brick structure in a nondescript neighborhood. Within the compound wall, the main building was surrounded by parking lots and various support structures: a morgue, a hospital jail, a police station, a maintenance shed.

After following Iman through the reception booth, I circled outside the hospital toward the director's door, where I hoped to bargain for my wife's leave. I passed an eight-foot-high picture of Saddam, under glass on the outer wall. Astride a white horse, long saber in hand, he was shown playing out one of his favorite fantasies, the "liberation" of Jerusalem. At the lower-right corner, an Iraqi woman followed her president while holding—what else?—a portrait of Saddam in a business suit.

Saddam could never get enough of Saddam. It was the rest of us who got tired of him.

I found a grim scene by the morgue. The corpses had piled up outside on the dirt, twenty-five of them in four neat rows. There were some in wooden coffins, others with only a tarp for cover. Parents and children milled about, slapping themselves in their grief. They could not take home the bodies until the doctors issued death certificates. The doctors were too busy with the dying to handle paperwork for the dead. So the rows kept getting longer.

Inside, the hospital was crawling with armed men in fatigues: army officers in khaki, Baathists in olive green, fedayeen in their black ninja-style dress, with hoods hiding all but their eyes. At the start of the war, Ali Hassan al-Majid had established military staging centers in a number of hospitals, which were less likely to be shelled.

The director was circulating upstairs, an aide told me. I went to the elevator and pressed the button for the second floor—maybe I would find him there. Anyway, I wanted to sneak a look at the "prominent" patient. It was an old attorney's habit of mine, to fish for the facts. In this case, some would call it a reckless curiosity.

The patient corridors were in bedlam. With all the rooms full, the wounded had been dumped on stretchers and gurneys. Most were injured soldiers, but I saw civilians, too, including several women and children. I saw men with missing legs, with open chest wounds, with visible holes in their heads—all moaning for help that did not come. One woman screamed as her husband bled freely from the stump of a lost hand. No one paid attention.

Hanging back outside the elevator, where I'd be screened by the comings and goings, I peered down to the cardiac unit, thirty feet away. Its door was guarded by two plainclothes fedayeen. I knew who they were by the oversize Series 16s—the best handguns available in Iraq—stuck in their belts.

As I loitered for the next half hour, I saw a number of peculiar things. No flowers or gifts arrived at the guarded door, or a single visitor. Odder still, doctors and nurses slammed the door as they went in and out. They shouted instructions to one another. This was not the kid-glove treatment that high Baath officials demanded.

By the time they served lunch, I *knew* that it wasn't a VIP. A guard stopped the woman wheeling the meal cart and took a tray inside. This patient was eating the same awful food as everyone else.

I went back to Iman on the fourth floor, where she worked in the kidney dialysis unit. "No sign of the director," I said. "By the way, have you heard anything more about that cardiac patient?"

After checking at the nurses' station, Iman told me, "Now they say the man up there is a prisoner of war."

"American?"

"Or British, they don't know." Iman frowned at me. "Why are you asking me so many questions? This is none of your concern." A POW was sensitive business, and my wife did not want me involved.

I thanked her and hiked down the hallway to Hamida's office. A doctor, I reasoned, might be privy to more details.

"Let me double-check for you," she said. When she returned, she, too, was frowning. "It is true, it's a POW," she said. "An American."

"Thank you, Hamida."

"There is something more—it is a woman." My sister-in-law must have seen something in my face she didn't like. "But why are you so interested?"

By then she was talking to my back. As I returned to my post by the second-floor elevator, I chanced upon a shift change at the cardiac unit, a new pair of guards. One of the replacements, a hulking man with a black crew cut and a Fu Manchu mustache, soon wandered off after his girlfriend, a nurse. The other, less impressive, sat in a chair by the door, as still as a stone. After a time I saw that he must be snoozing. I'd seen no one else entering the unit. This was my chance.

I eased into the flow of nurses and orderlies, stopping at arm's length from the guard. His head was back, his mouth open. The door to the unit was slightly ajar. After looking both ways for the Fu Manchu, I clasped the doorknob and pushed through.

The next twenty seconds would etch into my memory. I was in a large, sunlit chamber with supply cabinets and an EKG machine. A row of empty beds lay straight ahead. There were three private rooms to my right, three more to my left. Each had a glass panel, roughly two feet square, set into its wooden door at eye level. All of the rooms were dark—except for one, in the far right corner.

I took two steps forward to see inside that room.

I registered a well-built man, easily six-foot-three, standing with his back to me. He was all in black, except for the glimmer of his gold epaulets: a fedayeen officer. I could see his face in profile: dark skin, thick hair, broad mustache like Saddam's. His left foot was propped on a chair, his arm rest-ing on his knee. He was leaning out over the foot of the bed, menacing a pale figure who now stole my attention.

I cannot say how I had pictured this American POW, but I never imagined her as quite so small or quite so young. She was a child, really. Her bed had been raised to a forty-five-degree angle and faced me head-on, so I could see her clearly. She was mostly covered by a white blanket. Her forehead was bandaged. Her mouth was knit in pain.

There were two other men in the room: a note taker with a double chin, who sat near the officer, and a skinny translator with eyeglasses, nearer the prisoner.

I saw the officer look to the translator. Whatever he heard did not please him. Swinging from the shoulder, he slapped his captive with the palm of his right hand, then with the back of it. Her head jerked back and forth; she was a poor match for him.

I could not hear the sound of those slaps, but I felt them at my core. My heart was cut. In that flash of violence I did not see an American, some captured combatant. I saw a helpless young girl—someone's daughter, someone's sister. I saw someone facing death here, day after hellish day.

In that moment I felt compelled to help that person in the hospital bed. I had no idea of what I could do, but I knew that I had to do something. I had witnessed far more lurid crimes since the war began, but this one moved me in a different way. It was not some fait accompli, a body dragged through a crowd. This girl's future still held possibilities. And because I had trespassed into her life, all by myself, I felt strangely tied to her. She was my responsibility now.

It all happened so quickly. Minutes earlier, I'd been an ordinary man. I'd wanted only to escape a battleground with my family intact. But I now I had seen something not meant to be seen, and it turned me upside down.

Fearful that the translator might glance up and spot me, I withdrew to the outer door. I was about to open it when I heard some heavy footsteps, a croaking voice: "Haven't you had enough sleep?" It was the Fu Manchu, returned to his post, rousing his partner.

There was no other way out. I was trapped.

I took a deep breath and opened the door, greeting the guards as old buddies: "*Salaam Aleikem!* Have you seen Dr. Hamida?"

The Fu Manchu growled, "How the hell did you get in there? Who is this Dr. Hamida?" He grabbed a handful of my shirtfront, which gave me a good view of his bulbous nose, swollen mouth, and protruding dark eyes. The sleeves of his white dress shirt were rolled up to set off his muscles and tattoos: a dagger on his left forearm, a cobra on his right triceps.

The fedayee tightened his grip and said, "You better give me your ID." I offered my guild card, which made some small impression: "So you're a lawyer." He took his hands off me. "Who did you say you were coming for?"

"Dr. Hamida."

"Who's her father?"

"Ravi." (In Iraq, people are known not by surnames, but by their fathers or firstborn children.)

The Fu Manchu said, "We'll find out about this." He stopped an orderly and demanded, "Do we have a doctor called Hamida Ravi in this hospital?"

The orderly not only confirmed this fact, but said hello to me and shook my hand. After eight years with Iman, I was a familiar sight around Saddam Hospital.

The guard said, "You know him?"

"Yes, I know him. He is a relative of Dr. Hamida."

The orderly moved on. The Fu Manchu returned his

attention to me: "If this guard was sleeping, why didn't you wake him up and ask if your doctor was in there?" The sleeper scowled at me; I had made him look bad.

I said, "I did not pay attention to him. I just went in."

The Fu Manchu said, "Here, I'll teach you how to ask next time." He shoved me down the corridor with a crisp blow from his pistol to the back of my head.

As I staggered off, seeing stars, I felt more resolved than ever. But there was little I could do on my own with these guards on duty. I needed someone on the inside, like Hamida, to help devise a plan. I wanted to move fast, before something happened to the girl.

Hamida was not to be found. I tracked down Iman's supervisor and worked out a three-day leave for her. My head throbbed. Maybe I needed to rest and regroup.

On our way to pick up Abir, Iman asked why I kept rubbing my scalp. I grinned at her and said it was nothing.

Our neighborhood was an upscale district laced by broad boulevards and grassy medians with date palms. Half a block from our house, we found people gathered in mourning. Three cars were parked on the street with coffins on their roofs, ready for burial. *The disasters keep coming closer,* I thought.

A friend named Khaled gave us the sad news. A tailor called Abu Taha, a good man who lived seven houses down, heard that his elder son was held prisoner by the Marines. The boy was a simple soldier, nothing more, and the tailor went off to look for him. He took his wife and younger son because they thought the Americans would not shoot at a woman on board.

That was not logical, of course. All the Marines could see was an unknown vehicle bearing down on them. They opened fire before Abu Taha could say a word. The three bodies remained in the car until the next day, when they began to smell. The Marines dropped them near the Iraqi positions.

Abir was full of questions as we went into our house: "Why did the Americans kill Um Taha and Abu Taha?"

I said, "They thought they were enemies, so they shot them."

"When they come here, are they going to kill us, too? Will they kill my mother? Will they kill you?"

"No, no," I said, "they will not kill us. They are coming to help us."

At one o'clock I asked Iman to fix lunch, though I knew I could not eat. I went to my study. Three years earlier, when I decided it was pointless to practice law in a lawless land, I moved my office furniture here. My favorite place to relax was a leather executive chair with walnut trim. When I needed to *think,* however, I sat on a bare wooden stool, at a small bar by the window.

Today the window was blown out, and the street was empty. Dark smoke billowed from a building behind our house. I pulled the curtains, lay down on the couch, propped a book over my face. It was easy to feel pity for the POW. Her prospects were all bad. She might die from poor care, or from the interrogation. The regime might use her as a human shield, or simply dispose of her when it saw the war was lost. Unless someone interceded, the girl would certainly die.

Yes, it was easy to feel pity. But the next step, that was hard. The front lines in Nasiriya were changing by the day, the hour. I knew the Marines were based some miles to the south and east of the city, but getting to them would take a miracle. To drive would be suicide—along the route I'd be taking, any car moving toward the Americans could be targeted by the Iraqi army. To walk would mean hours of dodging the fedayeen, and hours more in no-man's-land, where I might be shelled from both sides. Even if I reached the Marines and convinced them not to shoot me, there was no guarantee they would trust my story about the girl. I might do all of this for nothing.

I thought of Hind and Abu Taha—if I believed in bad omens, I had them in bushels.

I thought of Iman and Abir—how would it be for them if I did not come back? Who would take care of them?

I went to my bath thinking I would not go—I had come to my senses just in time. But when I shut my eyes to relax, I saw the girl's strained face, and the man in black, and the blows that jerked her head again, and again . . .

Some might say I'd lived a sheltered life, but there was no real shelter in Iraq. Long before this latest conflict, the casualties had piled at my door. The regime had killed my favorite teacher, and the cousin I idolized. They'd come close to killing me more than once—for the crimes of speaking my mind and wanting to learn about the world.

Abir was a casualty, too, from the time she was two years old. They had violated my daughter so obscenely, so unspeakably, that they had made her father a dangerous man.

A man who might do almost anything.

Of all the cruelties I had witnessed in Iraq, I had intervened in a few, a handful. Most I had let pass, as there was little to be gained. But the POW was different. I had a chance to help, because Saddam's days on top of us were numbered. I could bring the Americans back to this girl before it was too late.

As I dressed in my study, my choices seemed clear. I had told Iman that I'd be driving out to Baba's to check on him. The car keys were sitting where I left them, on the TV in the family room. I could take them and go to the garage and live my life, and in time I would probably forget about the girl.

Or I could walk out our front door and into the howling storm for someone I had never met.

I tucked my denim shirt into my jeans. As I stared at my reflection in the floor-length mirror, something occurred to me. If I pretended this morning never happened, I would always be a small man. But if I seized my chance and went to the Marines, I would be a big man—in my own eyes—for the rest of my days.

This once, I would say no to Saddam. He could not have this girl; I would not allow it.

This once, my heart won over my head.

I left my study for the family room, keeping my back to the table where my wife and daughter ate their chicken and rice. I dared not look at them. I couldn't afford to tell Iman of my plan, for her protection and mine. Plus I knew that we'd have an argument, and I might not be strong enough to stick to my decision. Already I felt shaky. Fedayeen were massing like wolves on the streets. How could I possibly reach the Americans?

Iman called out, "Do you want us to go with you to Baba's?"

"No, you stay here," I said. I could hear my voice quavering—could she? "You get dinner together, and I'll come back as soon as I can."

It was time. I had far to go before dark. I moved past the car keys without breaking stride, and I walked out the door.

PART TWO

CHAPTER FOUR

M Y FATHER'S FATHER, Rehaief al-Rubaei, was a sheikh—the leader of a tribe of landowners— south of Nasiriya. His farm was the biggest for miles around, even after the government seized part of it for a railway. Al-Abthar, as they called it, was a fertile tract of marshland along the Euphrates River valley, irrigated by canals. Fifty workers harvested rice and wheat and barley, tomatoes and cucumbers, oranges and cantaloupes and pomegranates. There were horses for the overseers and cattle for the family's milk and beef, and for ritual slaughters on feast days.

Like the rest of southern Iraq, al-Abthar was best known for its stands of Khastawi date palms, which yielded both fruit and molasses. These dates, as long as your index finger, were known throughout the world as the very finest. Some were bought by the state at controlled prices for export. The rest went to market, five miles away, either straight to the merchants or through the tenant farmers who tilled our land.

Rehaief al-Rubaei, my *jadi*, was a formidable man, from his long beard and bristling eyebrows to his booming voice.

Even into his eighties, when heart trouble slowed him, he was called—respectfully—Wahsh, the Beast. Though Jadi treated me lovingly, I felt timid around him. He had that kind of presence.

Maybe I had heard too many stories, like the one about the man in their tribe with two wives. The first wife was barren; the other bore the man several children. The childless one became so jealous that she smothered one of the other's children with a pillow. According to tradition, the people brought the murderess to Jadi before carting her off to the police. He determined the facts and then meted out his own punishment, beating the woman with a palm branch as everyone cheered. It might sound barbarous today, but the people expected their sheikh to enforce the age-old code. They had no community without it.

Jadi himself had three wives, one fewer than the maximum allowed by the Koran. The first had several stillbirths. The second wife had a son, who drowned in the Euphrates at the age of eight. In her grief she fell ill, and died a few years later.

In Arab society, where land and wealth passed through the male line, Jadi had a problem. He found his solution unexpectedly, in the home of a neighbor who'd disputed their border for many years. This quarrelsome man had a single child, a girl. By her late teens, she had grown into a vibrant young woman.

"Why should we keep fighting?" said the neighbor, who hoped to secure his daughter's future. "If you marry my Zahra, all the land will be kept between us." The deal was struck.

Zahra al-Zangana, my *jaditi*, was warm and sweet-natured, if vain about the fine hair she wore in braids to her waist. She had blue eyes and fair skin, with the beauty marks prized among her people. She was also fifteen years younger than Jadi, who sensed that his luck had changed. Before long they had a son, Ahyal, and then a second son, my father: Odeh Rehaief al-Rubaei. Two daughters followed, conjoined twins who died in infancy. Still, the family line was assured.

To accommodate his growing brood, Jadi expanded his living quarters. He'd begun with a simple home of dried mud. As he added more wives, he built on wings made of brick, until the house became a horseshoe with an open courtyard in the Arab style. (Each wife had her own kitchen, of course.) There were barns for the livestock on either side, and storage sheds for the dates and grain in back.

The farm prospered, and the two boys thrived. All went well until my uncle Ahyal grew up and fell in love with a Sunni woman. After her clan rejected him, the young lovers pledged never to marry anyone else. Not long after that, Ahyal went out on his horse and failed to return. When Jadi found the broken body, he suspected foul play by the family that had spurned his son, though he had no proof. Some thought that Ahyal's horse had been spooked by a snake. In any case, my father was now an only child.

Amid such sorrows, Jadi stayed a compassionate man. After one of his workers died, he kept paying the man's salary to the widow and covered their son's education through medical school. The boy, Ibrahim, became a doctor in England; Jadi was very satisfied.

In an area with no hotels, my grandfather was famous for his hospitality. His guesthouse was built from bamboo, so well hewn that it never leaked in the rain. Jadi received his cronies there and entertained in high style. (As the Prophet commanded, out-of-towners could stay for up to three days.) When my father was nineteen, he prepared for five visitors by killing three sheep, which seemed to him more than ample. My grandfather berated him in front of the company: "If we have five guests, we must slaughter *five* sheep."

At the time, the scourge of Jadi's tribe was a wild gray boar. Along with its mate and a young one, it was loathed by all the farmers. The boar ate their crops and trod through their fields, and once killed a shepherdess and her child. The creature was renowned for its stealth; it eluded all hunters and attacked the careless ones. To make matters worse, the boar lived in a graveyard on my family's land, a wooded area said to be filled with the ghosts known as djinni. No one risked that graveyard at night—not since a neighbor had gone there seeking a lost cow. When his friends saw him next, he'd beheld something so horrific that his hair stood straight up. He was witless and deranged until the day he died.

Jadi was holding court one stormy night when a hen-pecked fellow said he had to leave because his wife might worry. Another guest, an older man, nodded solemnly and said, "Yes, we must listen to our fears, because God has put these feelings within us."

In the ensuing debate, Jadi took an extreme position: that a man should fear nothing, least of all from a wife. Feeling demeaned, the henpecked man said, "If you are not afraid, then why is this wild boar running loose?"

Jadi turned to my father and said, "Bring me my gun." When my father ran to get the shotgun from the main house, Zahra learned what her husband had in mind. She saw him privately and said, "Why do you want to do this? I have lost Ahyal. If something happens to you, there is no one to take care of me and Odeh."

But Jadi scorned her tears and said, "If I die, my son will know I was brave. Better for him to lose a father than to think his father feared a ghost."

Back at the guesthouse, Jadi vowed to the assembly that

he would kill the boar and they would celebrate with a sunrise breakfast. If he did not return by dawn, he said, they could come looking for him, but not before.

The others were certain he was marching to his death. They pleaded with him not to go, to wait for better weather. My father, an earnest young man, begged to hunt the boar himself. But Jadi was stubborn and fearless; he was the Beast. He ordered my father to stay and feed their guests. He put on his good camel's wool *juppa*—the Arab cloak—to shield him from the rain. Then he went off alone to face the boar and the djinni and whatever else might show its face that night.

As the legend is told, Jadi dug a trench in the graveyard and waited in the dark. When he heard a heavy thrashing in the brush, his ears guided his gun. There was a blood-curdling squeal; Jadi reloaded and shot again. Another squeal. He reloaded and shot a third time. Silence—one had escaped. He sat back in his trench for another chance.

Meanwhile, the guests blamed the henpecked man for making his senseless challenge. Some even cried. ("That is when I realized," Baba would recount to me, "that even though the people were afraid of your *jadi,* they still loved him.") They feared the worst after neighbors came to report gunshots from the vicinity of the graveyard. An hour passed, then another. Had the shotgun misfired? Had the boar overpowered him? The guests wanted to take their lanterns and search, but my father reminded them of Jadi's instructions. No one wanted to defy their sheikh.

It was just about dawn when they heard the rhythmic blasts of the shotgun. They rushed out to find Jadi, return-ing as promised, firing into the air in jubilation. As they ate their sunrise breakfast, Jadi saw that his son had burned his hand while serving the sauce for the mutton at dinner. He joked with his friends, saying, "I went out to kill the boar and nothing happened to me, but you let Odeh get hurt in the kitchen."

When they trooped to the graveyard, they found the bloody female boar and the baby, but the male was not there. Jadi was furious. "My gun betrayed me!" he said, and flung it to the ground. To calm him, his friends asked to hear about the djinni he'd encountered the night before. These tales were told with great gusto.

The boar was not seen in the graveyard or the fields again. A month after Jadi's adventure, they found what was left of it on the railway line, smashed by a passing train.

"And this," Baba told me, many years later, "is how you know the legend is true." He held up his right hand, still scarred from that famous night.

As time passed, Jadi relied on my father more and more to handle the farm's paperwork, collect rent from their tenants, and settle disputes within the tribe. Zahra was not content, however, for Odeh to remain a mere fellah, a bumpkin. She wanted him to be a doctor, but settled for a teachers' college in Baghdad when Baba was nearly thirty.

At school he discovered Najia Kazim, a slender young Shiite with a pear-shaped face and slate-colored eyes. Born to a small factory owner in the capital, she owned no veil or chador, and her dresses fell just below the knee. She was sensitive, emotional, sophisticated. My father had never met anyone like her.

Though Jadi might have arranged a more lucrative match, he was an indulgent father who denied his son nothing. Odeh would follow his heart. The wedding was delayed, however, after Jadi fell sick and my father dropped out of school to help with the farm. Two years later, when Najia graduated, they were married and settled into a three-story brick house in Nasiriya, with a two-car garage and a large garden. Two years after that, in 1961, my brother Ali was born.

Then came a run of bad luck—or, more precisely, of daughters. Batul was born in 1962, then Fatma in 1964. By the time Zayneb came along, in 1968, Baba was worried. When he died, all those daughters would drain the family's wealth into other clans, even if they received only half a son's inheritance, as according to law. And—God forbid—if something should happen to Ali, his lineage would end.

My father gave my mother an ultimatum. If their next child was a daughter, he would take a second wife.

During the next pregnancy, Baba moved his family to Najaf, into an apartment over a pickling plant, which Jadi had bought for the farm's cucumbers. My father planned to keep the place in Nasiriya for his second wife. He could shuttle between the two households, an hour's drive. As for Najia, she would live out her days in a second-rate flat that smelled of pickles.

You can imagine, then, how grateful my mother was to see me arrive on September 15, 1969. A year later, Baba left Najaf and moved us all back to the gracious house in Nasiriya, my home for the next twenty-five years.

As fate would have it, three more boys followed, my brothers Ahmed, Hussen, and Hassan. Baba was content. He always cherished my mother, even in those hard, daughterly times. But she would not let him off so easily. When he told her he loved her, she'd reply, "No, I do not believe you, because you were going to get rid of me if Mohammed was a girl!"

It became a joke in our family, part of our lore.

Owing to the circumstances of my birth, I had a special place in my mother's heart. I even had my special name for her—where my siblings called her Mama, I knew her as al-

Hbeiba: *my dear.* I was small for my age, more delicate than my rough-and-tumble brothers, and my mother protected and favored me.

For example: My first ambition was to be a concert pianist. My mother saw to it that I had a toy piano, which I hammered at incessantly. Baba hated the racket nearly as much as Ahmed, my roommate. My mother's solution was to move me into my own bedroom, and make Ahmed bunk with Hussen.

My brothers resented me, but what could they do? I had al-Hbeiba in my corner.

At school I learned to clap for Saddam, but first grade confounded me. Little things threw me off balance. When we went to our alphabet books, I did well until the teacher said, "*H* is for 'Hussein.'" That didn't seem right to me; it didn't match the picture. I was still wondering what had happened when she improvised again: "*S* is for 'Saddam.'"

"Austada," I called out, "we don't have this in the book."

Fed up with me, sweet *austada* struck me twice with her ruler across each palm. I was a soft little kid, and the spanking raised red welts. They hurt so much that I couldn't carry my book bag in my hand. I had to balance it on my shoulder all the way home.

I *hated* school, I told my mother. I hated the class and I hated the teacher. I hated the Baath national anthem that we sang each Thursday before our weekend break: *A homeland extends its wings to its horizon . . .*

Before entering my classroom one morning, I splattered mud on my pants. Thinking I had fallen and hurt myself, the teacher told me to go home. The next day I tried the same ploy—but this time she spotted me through the window. I had to sit at my desk all day in muddy pants.

Since the day of the caning, al-Hbeiba had decided that I was not ready for public education. She coached me,

"Don't tell your father that you don't want to go—he will only force you. Better tell him that you feel ill."

So I did, and never quite recovered. I stayed home the entire year and it was glorious, much cozier than first grade. My mother made heavenly lunches: stuffed grape leaves, pizza on round flat bread with chicken, a sweet-noodle *halawa*. When I tired of TV, I asked her to read me a story-book. My favorites were from America, where the people were generous, kind, and free to do as they liked. With al-Hbeiba translating as she went along, I was enchanted by these tales of faraway places.

Though he let his wife do as she pleased with me, Baba shook his head at my pampering. Like Jadi, he was a strong-featured man with big hands, a loud voice, and a look that stopped us in our tracks. His principles were unbending. One day, when I was eight years old, I picked up the ringing phone while Baba was trying to nap. My mother said, "Tell the man your father is not home." I did as I was told, and turned to find Baba scolding my mother. "Mohammed is still very young," he said. "If you teach him to lie now, he will always lie."

When I bickered with Ahmed or Hussen, big brother Ali would settle the fuss. But when one of us went too far, Baba would convene court in his handsome study, grilling us one by one to ferret out the truth. He was a gifted detective. After some culprit with an air gun pelted the gardener's bottom from the second floor, Baba narrowed the suspects to myself and Hassan. (The rest could prove that they'd been downstairs at the time.) When I swore that I'd been sleeping, Baba went up to feel my mattress. It was warm; I was cleared. His case hopeless, Hassan con-fessed. He drew the standard sentence: six lashes on the soles of his feet.

Such discipline was toughest on our tenderhearted

mother. She cried at movies; she wept bitterly upon the death of a neighbor woman she scarcely knew, out of sympathy for the children. As Baba dished out a spanking, she'd fret outside the study door, cringing at the sound of the switch. When the guilty one emerged in tears, she would hug him and stroke his head, console him with pistachios and sugar-dusted almonds.

While she spoiled us shamelessly at home, my mother would not stand for our flaunting our wealth outside. After Ahmed broke her rule one day by wolfing a banana down the block, she made him bring out a bowl of fruit to share. My mother had underrated my brother's greed; he wound up selling the fruit to the other boys, instead. When al-Hbeiba found the extra change in Ahmed's pockets, far beyond our daily allowance of one dinar, she marched him out the door to return it.

Unfortunately for Ahmed, my father also caught wind of his small enterprise. A spanking followed. My mother took the fault as her own, and she cried and cried. Then she drowned Ahmed in sweets and kisses. In our house, sins were not merely forgiven—they were rewarded.

CHAPTER FIVE

T HERE IS NO SHAME in being poor," my father would tell us. "Thank God that your fate is to live in a nice house, in a family with enough money. But you belong to the poor as much as to the rich. You must give them their due. Give to others, and God will give to you."

Like every good Muslim, Baba gave a percentage of his annual income from the family businesses to the mosque, for distribution to those in need. He would also bring livestock from the farm—usually a lamb or goat, on holidays a calf—to be slaughtered on a grassy patch between our house and the street. The animal was drained of blood and hung on a hook, its skin inflated and flayed for a rug. After the meat was butchered, a share went to the mosque and a share to the neighbors, but we were also generous with any passerby who paused by our door.

We were children of the modern age, and each of our bedrooms had its own television. At first they were black-and-white, but once Baba bought the first color set, we all wanted one. The old sets went into a closet. When larger-screen models came out, we wanted them, too. Though my father grumbled about the extravagance, he coddled us just the same.

I loved television, from Sinbad cartoons to the dubbed American action movies we played on the VCR. My favorites were Westerns with John Wayne. I liked the way he beat every villain to the draw, but also how he stood up for the weak. He was a hero with a big heart.

When a neighbor's family came for dinner one night, their children were glued to the TV as soon as the plates were cleared. As the man and his wife stood to leave, Baba asked whether they had a set at home.

The man said, "We have one, but it's not working."

Without another word, my father took a color set from the closet, loaded it in his car, drove the family home, and made it his gift. When he got back, my mother said, "Why didn't you give one of the black-and-white ones?"

And Baba said, "Why give something that is not the best? If we want to do good, we must give the best we can."

In Baba's view, a good Muslim "lived by his hands and his tongue." Deeds counted more than words; one must love one's neighbor as oneself. Above all, a Muslim should never harm others. Baba scorned the Ayatollah Khomeini, who brought the Shiites to power in Iran when I was nine years old and set off a catastrophic border war with Iraq. A true holy man, Baba said, would not sacrifice human lives over such trivial disagreements. Shiites, Sunnis, even Christians like Rafael, the goldsmith who often dined at our house—all were "children under God," he'd tell us.

While Baba was no fundamentalist, he was dutiful. He prayed five times a day and went to the mosque—unlike al-Hbeiba, who saved her energy for home and children. Many times my father made the pilgrimage to Najaf, Iraq's holiest city. When I was eleven, he set off on the hajj to Mecca. (From then on, he would be known publicly as Hajji, to acknowledge his devotion.) We eagerly awaited Baba's return three weeks later, knowing he would bear gifts. We were not disappointed. The Chevrolet was packed to its rooftop with the latest goods from Saudi Arabia, including a bicycle for me.

The one ritual that all grown-ups followed, even al-Hbeiba, was the Ramadan fast. For an entire month (a lifetime, it seemed to me), no food or drink was permitted during daylight hours. Though children were not expected to fast before puberty, Baba tried to inspire us with a dinar-and-a-half "bonus" for each day we held firm. When I was young and the whole world seemed a bazaar, it was an irresistible lure.

But that money was hard-earned. As skinny as I was, I had a robust appetite, and by late afternoon I thought of nothing but the cavern in my stomach. Halfway into Ramadan, I began raiding the kitchen, slyly choosing items that would not be missed at the evening meal—an apple or a cookie, say, rather than a slice of the fresh-baked chocolate cake.

My ruse was up when the maid found the wrappers from my Nestlé's biscuits. I knew I was in trouble when Baba called me into his study. He asked me, "How many days have you fasted?"

Choosing a number that might sound reasonable, I said, "Ten days."

Baba was all fatherly concern. "How are you managing? When you eat at night, are you satisfied? Do you feel hungry during the day?"

I stammered, "Yes, I'm managing."

Baba said, "Have you *really* fasted all ten days?" He stared at me with his big chestnut eyes, until I knew that he could see into my lying soul. Nerves are the enemy of deceit. I admitted I had broken the fast.

And Baba said, "Yes, I know." *How did he know?* His omniscience never failed to amaze me. "It is good that you told me," he continued. "If you had lied, I would have spanked you. Look, if you are not fasting, go ahead and eat. You don't have to sneak off to your room with those cookies. Go and have a good meal."

That day I stuffed myself in a bittersweet feast, mourning my lost Ramadan riches. My poverty did not last, however. The next day al-Hbeiba began slipping me extra coins to make up for my bonus. It would be our secret.

To this day, I have yet to fast through Ramadan.

Though I liked any game with a ball, I was the smallest boy in the second grade and it was hard for me to keep up. My puniness was a great burden. In my fantasy life, I saw myself as Bruce Lee or Jackie Chan, my movie heroes. In the school gymnasium, while others played at soccer or basketball, I attacked the heavy bag with special kicks and punches: *The Tiger,* or *Enter the Dragon.* When our new gym teacher, a kindly man, saw me at work, he asked me where I had learned such moves. Then he wrote down the name of a friend of his. "He will train you," the teacher said, "but you'll have to pay him."

I went first to my mother, who agreed to the class as long as my grades held up. My father was surprisingly enthusiastic; he must have worried about his fragile second son. He took me the very next day to Mr. Alaa, signed a waiver, and paid one month's fee—five dinars—in advance.

At my first session in kung fu, I could barely contain myself. Mr. Alaa's gym was split into two sections, a ring for the matches and a training area with punching bags, mirrors, and a rope-and-pulley contraption to stretch one's legs. There was a wooden skeleton and an anatomy poster

with all the muscle groups. Tacked on the wall were THE
TEN COMMANDMENTS OF KUNG FU:

1. Never start a fight unless you are attacked.
2. Allow mercy no place in your heart.
3. Win with speed above all.
4. Know that a bone gets stronger when a blow does
   not break it.
5. Surprise your foes to gain the key to victory.
6. Master every situation with spontaneity.
7. Strike first to win half the battle.
8. Make each part of your body an eye.
9. Always fight with a clear head.
10. Sweat more in training, shed less blood in the
    match.

There were nine beginners that day, all in a line in our
miniature belted suits. Mr. Alaa, a sturdy young man in a
black-skirted uniform, faced us without a smile and said,
"This is a game for the brave. Had you not been brave, you
would be at tennis or swimming, instead. You are heroes!
And I am going to be hard on you for your own good."

Then he said, "Just because you are here does not mean
you've been accepted in the class. I have to test you." It
was daunting, Mr. Alaa's test. (I would one day use it on
my own students.) He pointed to a place near the changing room, where we'd find a first-aid kit. "I will call on you
one at a time," he said. "If you don't want to take the test,
change your clothes and go home. If you want to take the
test, get a cotton pad from the kit and come to me in the
bathroom. I am going to punch you in the face, and you
will need the pad for your nose, for the blood."

Nine children inhaled as one. I was second in line, and I
didn't want anyone to say I was a coward. I remembered

what Baba used to tell me on the farm: "The person who never falls off the horse won't become a good rider." If I didn't get hit now, it would happen another day. I might as well get it over with.

Mr. Alaa called for the first boy, Namir. Years later he would be a kung fu champion for all Iraq. But then he was just a nine-year-old boy, like me, and it was a sad sight to watch him tramp to the first-aid kit and then across the gym to the bathroom.

Two minutes later, Mr. Alaa came out, wiping his bloody hand on a paper towel. Where was Namir? I imagined him passed out on the cold bathroom tile, his nose broken in five places. I envisioned what would happen when I came home in such poor shape. My father would spank me because I should have known better and left the gym. My mother would cry. My kung fu career would end before it started.

"Mohammed!" Mr. Alaa faced me. "Are you ready? If you're afraid, you still have time to leave."

"No, I'm ready."

"Then come with me." I trotted to get my cotton pad and followed Mr. Alaa into the bathroom. I saw no puddle of blood. Namir was in the shower, evidently conscious. I felt a tap on my shoulder—

"Congratulations," Mr. Alaa said. "You have a future; you are very brave. When Namir finishes washing up, you'll go next, to be ready for our training. Wait here until we finish the test." Then he went to a stall where he had stored a small pail of blood—or, more likely, some red dye. He stuck his right hand in the pail and left the bathroom as before, with a paper towel.

When it was over, three of our nine had changed and left, while the rest had passed the first of many tests. Over the next four years, I saw Mr. Alaa three times a week after school, two hours per session, missing only when I broke my leg in a practice match. For me, kung fu was the ideal sport. Size and raw strength mattered less than quickness, timing, and ingenuity. I became more self-confident at school. I stopped backing down.

Mr. Alaa became my friend and often drove me home. As we left the gym one day with three other boys, we met up with one of his older students, a teenage girl named May. "Sir! Sir!" she said breathlessly. "These men are after me!" On her way home, a young man had pinched her bottom, not knowing what he was in for. May knocked him flat. When two of his buddies joined the fight, she took off running.

By the time May had explained this to us, the three men arrived to confront her. Mr. Alaa said, "What's the problem?"

One of the men said, "We have something to settle with this girl," and they cursed and insulted her. After Mr. Alaa warned them to be more respectful, the first man said, "You step aside. This has nothing to do with you." Then he made a mistake. He tried to push past Mr. Alaa to get at May—and found himself flipped over in the street. He lay there moaning, in no hurry to get up.

The second man tried to throw a punch, but our teacher beat him to it. (*Strike first to win half the battle . . .*) After he fell, May kicked him in the face for good measure.

The third man pulled a knife. Alarmed, I picked up a rock and pitched it as hard as I could. My friends did the same. We missed the man with the knife, but somebody hit Mr. Alaa in the head. Unfazed, he disarmed the assailant and put him on the ground. The corner grocer called the police, and that was the end of it.

At our next class, Mr. Alaa came in sporting a bandage and said, "Who picked up the stones?" The other boys pointed to me, which seemed unfair, though it had been my idea. (I felt positive that the one who'd connected was

Hamad, since he was both nearsighted and cross-eyed.) Our teacher was unhappy, and not just about getting conked. We had doubted his prowess and had not used kung fu. He suspended us all for a week.

In Nasiriya, as in every city in Iraq, the local government was run by the Baathists. The top men, all Sunnis, funneled jobs and contracts to those who scratched their backs. While Baba must have bridled at this state of affairs, he was first and last a businessman, willing to work with whoever held office. His clan had flourished before Saddam was born, and it would be there after the man departed.

When an official came calling on behalf of some public monument, Baba gave with open hands, though he knew the money would line the collector's pocket. To assure the good graces of the top brass, my father held dinner parties in their honor. Every Ramadan, baskets of oranges and dates wended their way from our farm to the desks of the lazy and corrupt.

Baba would bend only so far, however. He refused to join the Baath Party, though it would have been good for business. He had liked Ahmed Hassan al-Bakr, a president "who loved his people," Baba said, until he died in 1982. (It was common knowledge that Saddam had his old mentor killed with an overdose of insulin.) But my father had no use for al-Bakr's successor. His grudge became personal after the regime took thirty-two of our best young Khastawi date palms—just pulled them up and replanted them by the main presidential palace.

Within the family, Baba might say, "This is not a head of state. This fellow is a thief!" When he heard about the latest Baath atrocity, he'd refer to Saddam as "God's vengeance on Iraq." When the news *really* set him off, he would call the president *makhboul*—an idiot, a fool. Our house was the only one on the block without a picture of Saddam on the wall. If a visitor noticed, Baba would say, "We have it in our bedroom, inside."

From what I could tell, most of my teachers were in the same boat. As civil servants, all of them had to be party members, but few seemed fervent. They went through the motions and stayed out of trouble, and lived to work another day. The exception was Mr. Jaleel, our adored gym instructor. I remember a rainy day in the second grade when the entire school was forced to join a march for the party. Seeing our long faces, Mr. Jaleel told our class, "I'll stand to the side of the street and pretend I don't see you, while you run home."

No one ran faster than I that day.

The seventies were a time of peace and prosperity for Iraq. Oil money gushed into the treasury, but not to Shiite cities like Nasiriya. When water dripped though one of our classroom ceilings, Mr. Jaleel was fuming: "Saddam swallows the money, and now we have leaking roofs!"

Such candor was rare. My schoolmates and I heard little about the political world except when the regime was drumming up the populace against Iran or Israel. Saddam was the center of our universe, a smothering presence. When a teacher came into the room, we were to leap to our feet and shout, "Long live Saddam!" In Islam class they called him "The Believer," or "One Who Fights for God." Even in mathematics, my best subject, we could not avoid him. When our teacher put an extra-challenging problem on the board, he would say, "This one is very hard. No one can solve it except Saddam."

Our school days ran from seven to twelve-thirty, when we were sent home for lunch before the heat of the day. To keep us going, the teachers let us out into the school-yard for a brief snack at eleven. I looked forward to those five free minutes all morning. I was always hungry, and my mother packed the best snacks in the class. In addition to fruit, she would make one of her specialties—like lamb's liver in a baguette, with lettuce, tomato, onion, and ketchup. Her sandwiches were so good that I had to watch that no one stole them from my bag.

In third grade I joined a Baath youth group called Talaia—because some friends were doing it, but mainly for the blue-and-yellow uniform. The shirtfront featured a full-color map of the Arab world, and my brothers were in awe when I came home in it. Al-Hbeiba said, "It's beautiful, Mohammed."

The next day I had a rude awakening. While the other students relaxed in their cool classrooms, I helped raise the Iraqi flag and marched up and down the courtyard for fifteen minutes, a long time in the Middle Eastern sun for a ten-year-old. I felt restless and fatigued; I was not cut out for this. On the third day, I turned in my uniform.

In adolescence I grew more subversive. I'd make fun of Saddam's crooked mouth, mimicking him before my friends. Or we'd find an empty classroom at recess, load spitballs onto rubber bands, and fire away at the president's picture. (Saddam's nose was the bull's-eye.) We must have thought we were immortal.

Around this time, they played (and played) a song on Iraqi radio called "You're Welcome, Father of Hallah." The father of Hallah was Saddam, and the singer—a stubby, bosomy woman named Maida Nassad—believed that her president was the finest thing since *baklawa*.

My friends and I despised this awful woman. We had the same opinion of our sixth-grade civics teacher, Majeeda, a Maida Nassad fanatic. She actually *looked* like Maida Nassad, with her short black hair combed in the same boxy style. When Maida appeared on a TV show, Majeeda observed what the singer had on and bought a similar dress—always at the height of fashion, an inch or two above the knee. (That was a daring statement in Nasiriya, where even a big-city woman like my mother wore an *abaya* out of doors.)

Majeeda's job was to deepen our patriotic feelings. She would drone on relentlessly about the life and glory of Saddam, his military exploits, his political wisdom. I got caned for sleeping through her lectures, and caned again for failing to memorize Saddam's speeches—I had a strong memory, but it was selective. I'd retaliate by sneaking in to draw chalkboard caricatures of our teacher, with enormous breasts.

My conflict with Majeeda came to a head the day she refused to let us out for snack time. After an excruciating lecture about patriotism, with five minutes left to dismissal, she said, "Would each of you please now draw a picture of Saddam?" Famished and grouchy as I was, my pen flew across the paper. I wanted only to get done and home.

Upon inspecting my scribble, Majeeda was outraged. My Saddam had one eye too big and one eye too small. His teeth badly needed a dentist's care. Most unforgivable of all, I had neglected to give him ears.

"Khabeeth!" our teacher hissed. *"Evil one!* I hope your parents will teach some respect to you!" Once again I was caned, along with two other modern artists.

When Baba saw the welts and heard what I had done, he echoed his warning: "Why do you act like that? We will all get into trouble." I shrugged off his advice until the day that a long black car with tinted windows pulled up to our school yard. Six security agents with guns and walkie-talkies sauntered through our recess and into the principal's office.

Five minutes later, they came out with Mr. Jaleel and marched him handcuffed through the yard. We watched in horror, and the older students knew he would not be back.

A few days later, the rumor raced through the school: he had been tortured to death.

Baba was distressed. This was oppression, he said. What the Baathists had done was bad enough. What made it *persecution* was that they'd done it in front of the children.

But that was the point, of course. My friends and I put away our spitballs and funny pictures and smart-aleck remarks. Our favorite teacher had given us a final lesson.

CHAPTER SIX

I LIVED HALF MY CHILDHOOD during the Iran-Iraq
War, which raged on and off through the eighties. In
the worst of it, when sirens seemed to sound at every sunset, we camped in the concrete shelter Baba had built
under the garden—not just my parents and the eight children, but the maid and the gardener, too.

By 1984, Iraq had lost so many young men that they
were drafting seventeen-year-olds. The war really hit home
for us when Ali was taken. Baba wanted to buy him a
deferment, but it was too late; his name had reached the
central ministry in Baghdad. For one whole year we lived
in drama. My mother scarcely slept. Each Sunday we
expected the ring of our doorbell, the flag-wrapped body,
the hateful words from some officious Baath functionary:
Congratulations, your son is a martyr.

The stars were with us. Ali landed a driver's job with an
officer who had no taste for the front. My big brother
came back in one piece.

One autumn, not long after school resumed in October,
a neighbor's house was hit by an Iranian Scud missile. All

three of their sons were killed, including my classmate Rassoul, a top student. They made a shrine at school the next day with pieces of the boys' clothing, but I was too sad to go. Not long after that, they shut down the schools and Baba took us to the farm for two months.

It stayed warm that year into December; I still remember the mosquitoes. But I loved al-Abthar in any season. In the summer, when the heat in Nasiriya might top 120 degrees, we could cool ourselves in the river or find some shady nook in the rambling old farmhouse. (The clothes cupboard made a fine hiding place, and came in handy after I bloodied a playmate's lip with a kung fu exhibition.)

At al-Abthar we caught *masgouf*, the river fish as long as a man's arm. We rode horses and picked fruit off the trees. I gorged on pomegranates till the juice ran down my shirt; al-Hbeiba fussed about the stains, but even then she would be smiling. Where the rest of the world seemed so shaky and dangerous, the farm was our one safe place.

Arabs keep dogs for security, not friendship. Our two mutts were brothers. There was a black one and a red one, but Baba considered them interchangeable and called them both Mashlouba. They lived in a little house by the garage to ward off intruders.

Nervous about rabies, Baba barred us from playing with the dogs, which meant that we had to wait until he left the house. For a child who missed so much school, they became reliable playmates. I would filch a chunk of steak from the fridge, tie it to a piece of cloth, throw it over a tree branch, and pull it up and down while the dogs jumped and barked. When I got bored with the game, I'd toss them the steak. To my mother's chagrin, I would also kiss those dogs. My penalty was a scrubbing in the bath.

One night we were awakened by a gunshot and piled out of the house. Baba had run off a pair of thieves and fired a parting bullet, apparently missing. My brothers and sisters and I were all excited . . . until we found the Mashlouba brothers, stiff and cold. The burglars had fed them poisoned meat. When the police hauled our loyal dogs away, we all cried, even Ali—even Baba.

From that night on, I was obsessed with the knowledge that my father owned a gun and knew how to use it, more or less. I suspected that he kept it in his study, off-limits to us. The door was always locked, the key safe in Baba's pocket. But one day I saw him leave for the bathroom and merely close the study's door behind him. I brazenly tiptoed in.

Until then I had come here to be questioned or spanked. Now, alone, I had the luxury of drinking the room in. I passed the towering bookshelves, packed with histories and my mother's school texts. I brushed my hand across the sleek walnut desk. I sat in the high-backed leather chair and slid open the desk drawer. There were Baba's two guns, one new and one old, along with an antique carved knife.

I stared at those weapons for several long, forbidden seconds. I knew that Baba was no John Wayne. He armed himself only when he went to the farm to pick up the cash receipts. (Leery of banks, he kept the money in a safe at home.) Yet there was something mysterious about those dull metal objects. Something scary . . .

Spooked, I shut the drawer and ran from the room.

We lived next door to Abu Ismael, a blacksmith who supported a large household, including his arthritic mother. Around noon one day we heard a terrible commotion and ran out to look. A gas canister for our neighbors' stove had caught fire, and the whole house was ablaze. With Abu

Ismael away at work, his wife found their children and ran from the conflagration—only to find that her mother-in-law was still inside. The old woman could barely walk. Surely she was done for.

To my astonishment, before anyone could say a word to stop him, Baba poured a pitcher of water over his body and dashed into the flames. My brothers and sisters and I could not believe it—was he mad? We thought we had seen the last of him.

A minute later, Baba came out carrying the old woman. She was petrified but unharmed. Though the house burned to the ground, no one was hurt but my father, whose ankles were singed above his sandals.

I cannot tell you how proud I felt. The entire neighborhood turned out, all my playmates. I was the son of a hero. I did a lot of bragging that day.

Baba's friends were mystified. They could not understand why my father, this shrewd and sober businessman, would act so impulsively. They asked him, "Why did you put yourself in such danger? After all, it was only an old woman."

My father told them, simply, "Because each life is precious."

I had heard him say this many times, but now I understood.

Raad was my favorite cousin, the one I looked up to the most. He was ten years older, the middle son of my mother's half brother, a wealthy textile merchant in Nasiriya. Raad wasn't large or muscular, but if he had something to say, he would say it to your face. He bowed to no one. Though he was too blunt to be popular, I always felt proud when relatives told me I looked just like him.

From Raad I learned about courage.

His mother was sickly, and my parents once tried to get Raad's father to take a second wife, even offered a house to lodge her in. Understandably, Raad's mother held a grudge. After my brother Hassan was born, she did not come to visit for several weeks. On her arrival she had an excuse: "I'm sorry we couldn't make it before, because Raad was sick."

Raad frowned at his mother and said, "Don't lie, I wasn't sick. You shouldn't be afraid of my uncle or aunt or anyone else. You should *never* be afraid."

I was six at the time; Raad was sixteen. They sent us out for groceries for dinner. When we brought our basket to the cashier, he was busy watching Saddam on TV. It was another speech, a long one. For several minutes the man ignored us.

Patience wasn't one of Raad's virtues. "We've been waiting forever—let us pay," he said. When the cashier failed to respond, Raad looked at Saddam and said, "You know, he's not talking about anything important. He's just talking."

The cashier glared at us and said, "Are you *crazy*, to say something like that? Just get out of here! I'm not selling anything to you."

Raad said, "So you won't sell me anything? Okay." He grabbed a jar of jam and hurled it at the TV screen. Saddam's image shattered into a thousand pieces. We walked out with the cashier cursing our ancestors.

Dumbfounded, I asked my cousin, "Why did you hit the TV instead of the cashier?"

Raad told me, "You are young, you don't understand. I didn't go there to steal. If I hit the cashier, they would say it was robbery."

And I said, "But it's more dangerous to hit Saddam, you know."

Raad smiled his magnetic smile and quoted a proverb: *Who puts this head here can take it off.* My own head spun at this idea—that no one but God held power over us, not even Saddam. It was earthshaking, like the first time I heard about atoms or outer space.

That day came back to me ten years later, when *I* was sixteen. We found out that my cousin had been working with al-Daawa, the Shiite terrorist group most feared by Saddam. The problem was his friend, Sadik, whose girlfriend was poison. She was waiting for him on the street one night when a party official tried to pick her up. When Sadik arrived, he told the intruder to leave them alone. The Baath man pulled a gun and said, "I could shoot you now and say that you tried to kill me, and nothing will happen to me." And he took the girl with him.

When Sadik saw his girlfriend a few days later, she taunted him—he was a coward, no man at all. Sadik, who'd been drinking heavily, got so agitated that he lost track of his words. He feared no one, he said. To prove it, he told the girl about al-Daawa and his secret work.

A few months later, Sadik fell in love with someone else and got engaged. Crazed with jealousy, his old girlfriend informed on him. (The regime rewarded her with a Toyota.) When Sadik was tortured, he named more than a dozen men, including Raad.

Hanging my cousin was not enough to make their point. They delivered his body to the family with his fingers cut off. His mother had a heart attack and was hospitalized.

When Baba came home from work the next day, he told us to get dressed to pay our respects. My uncle's house was quiet as we pulled to the curb, a tribute to the two security agents outside the front door. Their job was to screen visitors and limit the crowd, and to listen for signs of excessive grief. If they didn't like what they heard, they banged on the door and ordered the mourners to quiet down.

The regime was taking no chances, because they never knew what a case like this might trigger. Another young man from Nasiriya, a soldier named Ghazi, once came home on leave to find his brother executed. He took a can of gasoline to the nearest square, where Saddam's portrait hung in a wooden frame. Ghazi sprayed the portrait, then himself, and they burned together.

When no one is listening, people still call the place Ghazi Square.

"You can't give condolences here," a third security man told us through our car window. "Raad doesn't deserve them. He deserved to die." Baba tried to explain that it was just a visit, not a condolence call, but the agent could see my mother and sisters in black. They would have to change before we could go in.

I could take no more. From the backseat I said, "Why can't you let us inside? It's our family!" As soon as I started in, Baba cranked up the radio and closed the window so they would not hear me. Shifting into reverse, he caught me making fun of the security man's shaggy mustache.

I had never seen Baba so angry. He'd always honored the Arab tradition—that a son who turns sixteen becomes the father's friend, and can no longer be physically punished. But in that moment, for the first and only time in his life, Baba raised his hand to strike me in the face. Somehow he stopped himself. In anguish, he said, "Be quiet, Mohammed! Don't point at that man—he'll see you and come and get you!" Then he said, more softly, "And I can do nothing if he comes and gets you."

Minus our mourning clothes, we returned to my uncle's house. He led us into Raad's bedroom, where no one could hear us, and we all wept freely. I slipped a picture of Raad into my pocket. Back home I hid it in an album under my mattress.

I was different after that. Where the loss of Mr. Jaleel had quelled my public heresies about Saddam, now I stopped my snide comments even to friends. I was a long tongue no more, to Baba's relief. He gave me back the car keys. He worried less about each knock at his door.

But was I less of a threat? When Raad was killed, I saw that a person's life had no value in Iraq. Something hardened in me.

And I hated Saddam more than ever.

Having lost one year to a fake illness, another to a real one, and a third to the Iran-Iraq War, I turned eighteen before entering high school. As much as it pained her to see me go, al-Hbeiba sent me to live with her younger brother in the capital, where I could get a proper education. I was off to the big city.

I liked my uncle Mahmoud Khayoun, a utility executive and part-time journalist. Like most professionals, he backed Saddam, but without much enthusiasm. (He wasn't nearly as zealous as his older brother, my uncle Ali, an up-and-comer in the Baath Party.) With two daughters my age, Mahmoud approved of my teaching them kung fu, even after I accidentally broke one girl's finger.

Baghdad was a dynamic place in the late eighties for a young man with a little money in his wallet. Roaming the streets with new friends, I went to movies like *Rocky* and *The Godfather,* to restaurants and coffee shops. I joined my uncle's swim club, a luxury in a desert country. Summers returned me to my family at al-Abthar. Life seemed normal again.

But not for long. In August 1990, before the start of my senior year, Saddam invaded Kuwait. Even early on, when our country seemed to be winning, most people in Nasiriya were against the war. Many Shiites from the south had lived in Kuwait at one time or another. During the conflict with Iran, the Kuwaitis sent us food. What was the point of this aggression?

And after the U.S. and the international community came out against the regime, we knew the ending would be unpleasant.

Once again school was suspended, though I kept my student deferment. I spent much of that fall on the farm, helping with the ledgers. Business was bad. There were food shortages everywhere, even on the farm. Cheese and sugar were nowhere to be found, and meat was scarce. After the Americans and their allies launched Desert Storm the following January, things got worse. In Nasiriya we cut back on bread and apples. We slaughtered one of the farm's prize bulls, even a pregnant cow.

The allied bombardment was intense in our city. When they brought down Victory Bridge, our only pedestrian passage over the Euphrates, forty-seven people were killed and more than a hundred injured. One casualty was a thirty-year-old contractor named Amer, a close friend who did business with Baba. He was loyal and straightforward and honest and honorable. When the price of iron or cement blocks went up after he bought them, he still charged us the old price.

Amer was a striking man, with olive skin, a long nose, black eyes, and curly hair that he wore with pride. He had a fiancée named Widad, a teacher. When they couldn't find Amer's body, Widad visited the bomb site each day. Gradually she fell apart. She'd leave her house disheveled and wild-eyed. Her speech, once so proper, became slurred and nonsensical. When she saw an olive-skinned man on the street, she would run to him—"Amer!"—and kiss him.

She, too, lost her life at Victory Bridge.

CHAPTER SEVEN

S ADDAM CALLED IT "the time of treason." I called it a fiasco: the Shiite revolt of 1991.

It began with the Iraqi army's headlong retreat from Kuwait. On their way back to Baghdad through southern Iraq, the soldiers were met by civilians who seized their guns and tanks for a new war, this one against the regime. At the same time, exiled fundamentalists poured back over the Iranian border near Basra. These rebels had heard the first President Bush urge them "to take matters into their own hands and force Saddam Hussein, the dictator, to step aside."

At least that is what they thought they heard.

The uprising broke out in Nasiriya on March 2, the day after Bush called an end to the war, before spreading to a dozen other cities. I had never seen so many people downtown, tens of thousands in a frenzy, chanting their defiance: *Saddam, wherever you fly, we will catch you!* Some carried pictures of Khomeini, others of the ancient Shia imams, Ali and Hussen. They marched like men who thought they would soon be in Baghdad.

As the demonstration wore on, the people loosed their

pent-up anger. For decades they'd been ruled by a ring of thieves, and now they would turn the tables. No public building was safe that day in Nasiriya. Government offices were sacked, furniture smashed, documents destroyed. Schools were looted and burned, even hospitals. Ambulances were taken and repainted for private use; priceless lab equipment was wrecked. The vandals took fans and clocks, desks and phones, sinks and toilets.

The target of targets was the Governate, a columned structure that looked like a miniature version of the White House. The governor was hated by the poor for a recent wheat scandal. (After commandeering the grain set aside for the public, he sold it back to the market in Baghdad and doled out inferior flour to the hungry.) He was hated by the wealthy for extorting "donations"; he once came to our family's hotel with his hand out for the equivalent of $50,000. If you did not pay, you went to jail.

Now that haughty man would pay in return. The throng softened the building's defenses with rocket-propelled grenades, then broke in and killed the guards. They ransacked all they could carry, from Persian rugs to the gold fixtures from the governor's private bathroom.

I saw much of this on my way to our stores, where Baba had sent me to secure what I could. I began at the apparel shop, piling the pricier dresses into boxes and packing them into my car. Soon an older man with a flowing beard and a Kalashnikov came to assist me. I nodded my thanks, assuming that he knew our family. We moved on to the cosmetics store and worked as a two-man assembly line, passing cartons of perfumes into my trunk.

When we finished, the man stepped to the goldsmith's next door and said to me, "Okay, let's get a move on."

I said, "You can't go in there—that's not our store."

It seems we'd had a misunderstanding. The man had

presumed we were partners. Now he thought I was cheating him out of his share of the loot. To convince him otherwise, I showed him a photo from inside the cosmetics shop of me and Ahmed.

"Okay," he said, "but now I'm stuck with nothing. I want a reward."

That seemed fair to me. As I pulled out some banknotes, the man loaded his rifle and said, "I'll take all of those—and that, too." He pointed the gun at a box of perfumes nearing their expiration date. I was happy to hand them over.

The news came from a loudspeaker, strapped to a car's hood: *The governor will be executed tomorrow at five o'clock in the afternoon!* They had found him hiding at Saddam Hospital.

The next day I went to the main square on Haboubi Street. There were fewer people out, perhaps two thousand, all jostling for a better view. Many wore white robes with a picture of Imam Ali sewn over their hearts. Their turbans were green, the color of heaven in Islam.

I squeezed my way toward the center of things. A murmur passed through the square, heralding a pickup truck that parted the crowd. A bald man in pajamas, his wrists manacled, was seated in the back.

The governor was in a deplorable state. Like many from Tikrit, Saddam's hometown, he was tall and fair complexioned, but now his head was blotched from the blows of his Shiite guards. To express their contempt, they were using their sandals. He begged them to stop: "Don't kill me, please, for the love of God!"

As they took him down from the truck, the crowd spat on him. The guards forced him to his knees on a plastic tarp spread on the ground. A mullah came forward and told him, "*Now* you can rely on Saddam. We shall execute the decision of a just people."

God is great! the people cried.

A strong young man whistled his sword through the air. It struck the governor's forehead with a thud and a spurt of blood. I had to turn away.

God is great!

With the governor down, in his death throes, another man came and put two bullets into his head.

Today is our day of revenge!

The people were ecstatic, the mayhem only beginning. After they mutilated the local security chief, I could tolerate no more. I left before they dealt with the head Baath official, who'd thrown on a woman's *abaya* and handed $5,000 to his brother-in-law to drive him out of town. The crowd found him in the car's trunk, where the in-law had led them. They used a bulldozer to finish him off.

I detested these Baath men as much as the next; I had wanted to see what the people would do to them. But I had not wanted them killed so brutally. I felt uneasy. I sensed that we had all crossed a line, and that someone might be taking notes.

The Shiites held on to Nasiriya for several weeks, though no one was really in control. Private scores were settled, old resentments fanned. We became a city of predators, united only in our hatred of Saddam. When Baba protested the seizure of our rental house for a rebel headquarters, one of the young militants asked another, "How many people have I killed today?"

"Four," his friend said.

Pointing his gun at Baba, the first man said, "No problem if there are five."

I got my father out of there just in time. On the way home he sat glum and silent. What was there to say? Our new world seemed not much better than the old.

*   *   *

We learned the hard way that President Bush had no wish to get stuck in a civil war—or in a Shiite firestorm that might strengthen Iran. Soon enough, the regime made its comeback. As American pilots held their fire, Iraqi army helicopters crushed the Kurdish rebels in the north. Meanwhile, elite Republican Guard units set about cleansing the south of guerrillas or anyone who looked like one. When rockets weren't enough, they used mustard gas.

In Nasiriya it began with a callous shelling. Our hotel was hit. The guard there was killed, along with his nephew, a deserter seeking refuge. Then the soldiers came to town. Signs hung from the barrels of their tanks: LA SHIA BAAD AL-YOM.

No Shiites After Today.

Under the whip hand of Chemical Ali, they killed people right, left, and center. Any man in a black or green turban was presumed guilty. The soldiers would close off a block, bring out the young men, line them against the nearest wall, and shoot them.

A few managed to escape the net, like Raad's two brothers, who settled in Denmark. But many more were taken to Baghdad for interrogation, more than a thousand from our city alone. Most simply vanished with no word of their fate, except what I happened to learn from my uncle Ali Khayoun. His son-in-law also worked in intelligence, and they were debating whose system of torture was better—Special Secuity's, or the army's. To make his point, Ali noted that Special Security had dumped the Shiite rebels live into vats of sulfuric acid.

It was cleaner that way, my uncle said.

I knew many who disappeared. There was Maage Jabbar, the muezzin; Dr. Baghdadi, the psychiatrist; Abdul Aziz Menawer, the car dealer; Jafar Jeahz, the butcher. One day they were on the streets, smiling as you passed. The next day they were shadows who could not be mentioned in public.

For the rest of us, life got harder. Saddam punished the south by blacking out our power for weeks at a time—a measure he'd resort to, off and on, for the next twelve years. We used kerosene lanterns for light, charcoal stoves for heat. We lived like the people of Jadi's time.

When they cut off the water mains, we drank from the Euphrates, where bodies had been drifting since the bombing of Victory Bridge. Using a car battery, we'd boil the water in an electric pot and treat it with chlorine. The poor took their chances with the raw river. Dysentery was epidemic.

The shortages got worse. With wheat flour nowhere to be found, people visited the farm to offer exorbitant sums for 100-pound bags of barley—our animal feed.

Ever since the days of my storybooks, I had admired the United States for its humanity and culture. For me it had always stood for freedom and human rights. But after the Shiite uprising, I felt betrayed. America had looked the other way while Saddam cut us down. Where was the humanity in that?

Back in Baghdad for my last year of high school, I met a man who would leave his mark on me: Ciao Lee, a Shaolin kung fu master with an eighth-degree black belt. He was a small, weedy man in his mid-forties, nothing special to look at. But he had extraordinary speed and power. In one famous exhibition, ten weight lifters came at him at once—and he mowed them down like a scythe. His legs were a blur. He moved more quickly than anyone I had ever seen.

Ciao Lee had trained champions in Iraq. It was a great advantage to learn from a Chinese coach, and few were accepted into his class. I was a lowly blue belt, and Ciao Lee seemed unimpressed with my skills. "Who was your coach?" he said.

I said, "Mr. Alaa, in Nasiriya."

And he said, "Mr. Alaa must be incompetent."

My face flushed. I said, "No, he's not incompetent. Excuse me, but you cannot talk about my teacher like that." I had *done* it this time, I thought. I was bound to be sent packing.

Of the fifty-two candidates, Ciao Lee kept only fourteen: eight because they had less glaring flaws than the rest; four on orders from Iraqi intelligence; one, a Filipino woman, because she reminded the teacher of people from his homeland—and me, "because I criticized your coach and you were loyal to him."

For this man's students, kung fu was more than a sport. It was a driving force, a way to live one's life. I saw that I could never know enough. As Ciao Lee said, "Kung fu is a sea. No matter how much you learn, you will always be at the shore."

Through kung fu I met my first Jewish friend, a Baghdadi named Adam who later moved to Lebanon. Growing up, I knew nothing about the Jews. At school, where our teachers condemned Israel's treatment of the Palestinians, they were said to be colonialists, imperialists, racists. But Jadi had stories about the souk in the 1960s, where a Jewish market still did good business. Jews were able merchants, my grandfather said—no more, no less.

Baba said he had no problem personally with the Jews, but it was dangerous to mix with them because they were hated by the government. His warning made me more curious. Adam was short and blond, with a good heart. Once I broke my nose in training and went to the hospital for surgery, with my friend at my side. When the doctor said that I might need blood, Adam volunteered. His type was O-positive, like mine. Jews and Arabs, I supposed, were not so very different.

In school, my new favorite subject was English. Maha Ghazali, our teacher, was elegant, intelligent, and—as I would find—very bold. She was twenty-seven, only five years older than I. She had short blond hair, green eyes, a turned-up nose, and a delicate mouth. I imagined that women in Paris must look like this.

There was nothing in the air between us, not at the start. Then Maha began calling on me more often, and keeping me after class to discuss some point of grammar. I liked gazing at her from my desk. I looked forward to our talks. But I had no romantic strategy—this was my *teacher*. She lived in another realm.

En route to Nasiriya one weekend, my bus went off the road and rolled over. Though several students were injured, I got out without a scratch. When the news reached the school, Maha called Baba's house to check on me two days running. She got no answer because we had left for the farm, which had no phone. Picturing me in a coma, she found my address and hopped the train to Nasiriya. Our house was vacant except for the gardener, who drove her to the farm.

I was surprised to see her—had something happened at school? She told my parents that she'd needed some "important information" from me, and agreed to spend the night. We rode horses and had a good country dinner. When I drove her back to the station the next day, I said casually, "You haven't gotten the information you came for."

Sounding hurt, she said, "You are one of two things. Either you have no emotions, or you don't *want* to feel them."

Then she said, "I love you."

And I told her what now seemed obvious: "I love you, too." I had known this for some time, but it took Maha to break the barrier. She taught me to speak from my heart.

CHAPTER EIGHT

THE PROSPECT of getting drafted filled me with dread. Bad enough to follow the orders of Saddam's officers—to do God-knows-what in God-knows-where. Worse still to do all that marching and live in barracks with no air conditioning. I was a pampered kid, and the army scared me to death.

My lost years wound up saving me. By the time I finished high school, it was peacetime. With less call for cannon fodder, it was easier for Baba to buy my freedom with a check to the Ministry of Defense, a common practice among families who could afford it.

But my reprieve was short. It would be much harder to duck my duty to university.

When al-Hbeiba gave me a new piano for graduation, it renewed my childhood passion. I spent blissful hours playing Beethoven sonatas—what could be finer than a life spent at the keyboard? And if music did not pan out for me, I planned to become an actor. I had no desire to sit in a stuffy classroom and learn some workaday profession.

But Baba had another idea. Like many college dropouts,

he felt insecure about his résumé. He would never forget the time that he met my mother's uncle from Mosul, an eminent man with a doctorate in nuclear physics. It was shortly before my parents' wedding, and my mother begged Baba to fib, just this once—to tell the uncle that he had his bachelor's degree.

Two minutes after they met, the uncle asked him, "How did you do in college?"

And my father said, "I did not finish college." My mother almost burst into tears. (As Baba tried to explain to his humiliated fiancée, "If I lie today, someday your uncle would find me out. The rope made of lies is a short rope." She was furious just the same.)

Though Baba never got his degree, he became a missionary for higher education. He would say to me, "Let me ask you something. Look at the clothes you are wearing—how long will you have them before you'll want to buy some new ones?"

Assessing my well-worn shirt, I said, "Oh, a month or so."

"And what about the money in your pocket? How long will that last?"

"I'm going out tonight," I admitted, "and it will be gone by tomorrow."

"Exactly!" Baba said. "Nothing lasts forever, except your degree. It is shameful to boast about your car, or to say that your father owns this property or that one. You should brag about your degree, your achievement, your intellect!"

Besides Ali, who went to dental school, Batul had a master's degree in business management and Fatma taught French. Nothing less was expected of the second son. When I brought Baba my application for the music academy in Baghdad, he pushed it aside. "No," he said, "you are going to study law." Al-Hbeiba was bent on my becoming a judge, and Baba decided that I would not disappoint her.

In the autumn of 1991, I left my piano behind and entered the law college at the University of Basra, a four-year course.

Late in my first semester, a classmate from Nasiriya coaxed me into joining him on a weekend bus trip back home, where his brother was getting married. I agreed, though I wasn't high on the idea. The UN embargo had brought austerity and lawlessness to Iraq. Rings of highwaymen roamed the routes from the south. My friend and I were departing on the twenty-eighth of the month, payday, when workers and soldiers on leave traveled with full pockets.

As we mounted the bus on a cool, rainy morning, I admired my friend's good imported black suit, bought for the wedding. When I got back to Basra, I thought, I would buy one just like it.

Then I stopped short. A few minutes earlier, I'd left my briefcase on the seats behind the driver, to save our place. Now it was several rows to the rear. In its place was a sheikh in a turban and *juppa*. I said, "This is my seat."

The sheikh crossed his arms and said, "Once I sit down, I don't get up."

I said, "This is not a good way to behave, moving my stuff. You could ask me nicely if you want to sit here."

And the sheikh said, "I am not going to ask your permission. Do what you want, but I won't stand up. I am a Sharif, you know." That was the name of his clan. Sneering at my leather jacket and jeans, he added, "And I don't get up for men who dress like Westerners."

My friend tugged at me and said, "Forget it, let's just move. If you get into something with him, you'll have a whole tribe to answer to."

We moved to the middle of the bus, and I fell asleep as we rolled off. The next thing I knew, my head was bouncing off the seat back. The bus had stopped. Three masked men with machine guns jumped aboard. Outside we heard an eruption of gunfire. I peeked out and saw a dozen high-waymen surrounding us, shooting into the air.

We were an hour outside Basra, at a point known as Meat Hill. The bandits had waited around a bend, where they could not be seen before it was too late. They blocked the road with boulders and leaped out at the bus as it made the curve. Our driver knew how these men worked—there was no way out. If he backed up, the ones poised behind us would shoot him. If he pushed through on the shoulder, others up ahead would gun him down.

The men in the aisle began by clubbing the driver with their fists and guns. They wanted his fare box, but the main idea was to intimidate the passengers. It worked. We stampeded to the rear of the bus—all except for the sheikh who got up for no one. One bandit ripped off his turban and beat him about the head with it, until the sheikh scurried back with the rest of us. I broke out laughing. It is one of my big flaws—when I see something funny, I lose control, no matter the situation.

Brandishing grenades, the bandits told us to return to our original seats and empty our pockets. As they worked their way down the opposite side of the aisle, one with a gun and another with a bag for the loot, they made it clear that our pockets would not be enough. They would take our clothes; they would take anything that caught their eye. A few rows ahead of me, a pretty young woman took off her gold rings and bracelet and dropped them in the bag.

The bandit said, "We don't want that gold—we want *this* gold." He and his partner yanked the woman, screaming in terror, from her seat. Her husband, seated next to her, was paralyzed. He uttered not a word as they pulled her from the bus.

Behind me, a man slipped his ID card inside his shoe. He placed a handgun in the lap of the old woman next to him, asking her to hide it under her *abaya*. He must be Baath or security, I thought—more likely the latter, since he dressed in civilian clothes. Here was a bonus for the masked men. They would kill him for sure, add his gun to their stash, and make good use of his ID. I knew that his life was over as soon as his seatmate panicked and shouted, "This gun—*it's this man's gun!*" They grabbed the man greedily and took him off the bus.

Across the way, another old woman refused to give up her money. She told the bandits, "My son is with you."

"So you know who we are?"

"Yes, I know who you are," she said, unaware that she had sealed her fate. They shot her dead with a short spray from the machine gun and left her slumped in her seat. It was an awful thing to watch—to see men act like animals, with no glimmer of remorse.

As the bandits circled back along our side, they had no more trouble. They took my friend's suit, leaving him in a T-shirt and boxers. I had everything ready on the seat back tray: my watch and wallet, my cell phone, even my pen and sunglasses. The one with the gun said, "Take off your jacket and your shoes." They left me my jeans, and I made the mistake of smiling. With his free hand, the man with the gun punched me in the jaw, snarling, "So you think this is funny?"

The one at the door said, "Don't waste your time with him—bring him over here and get the rest of the money." I was lifted from my seat and propelled to the man at the door, who sent me reeling from the bus with a belt from the butt of his machine gun.

As I stumbled out, I saw the security agent sitting cross-legged with blood oozing from his head. The young woman sat next to him, not visibly harmed. I went up to the nearest bandit, hoping to reach some understanding. He slapped my face and told me to sit with the other two.

I shivered in my thin turtleneck shirt, and my jaw ached. Still, I told myself, I was in better shape than my companions. The agent was finished. The woman was a special case. The bandits might rough me up, but what reason would they have to do more?

As we sat, the other passengers were brought off the bus. Then the bandits told them to return to the door, one by one, for a frisking. Those who were clean, who had kept nothing, would be allowed back aboard. Of the thirty or so people remaining, four had money in their socks and failed the test: two older men, two younger. The bandits made the older men strip naked and walk a gauntlet, with the other passengers pummeling them as they passed. Any who held back, the outlaws promised, would go next.

The two younger men were taken off to a car. On the way they were bashed with guns until their faces were a bloody pulp. Was I next? I was used to getting my body beaten up in kung fu, but I wanted to keep my face.

I did not see those young men again.

With the rest back on the bus, a group of bandits approached the three of us on the ground. They were led by a large and fearsome man with one arm. He was all in black—his gown, his hair, the eyes that gleamed above his black scarf. He wore a black leather glove on his left hand, which held a small automatic gun. His men called him Abu Layl: Father of the Night.

He told us to get up. "Good hunting!" he said to his men. "Where is the security guy?" He looked straight in my direction.

"It's not me," I said, pointing to my right. "It's him!"

Abu Layl said, "Shut up!"

And I thought to myself, *Welcome to hell.*

The agent seemed befuddled, whether from the blow to his head or his professional training. (These people often diverted suspicion by acting clumsy in public.) Abu Layl turned to him and said, "God has brought you to me. You see, my right arm was taken by Saddam. When his people arrested me, they cut it off with a saw. So I will take my revenge on you."

The security man fell mute. He and his pals had everyone in Iraq quaking, they had tortured the whole world, but now he stood there tongue-tied. I felt only bitterness for him. Hoping to earn some goodwill, I told Abu Layl, "I'll deal with this guy for you," and kicked the agent in the head. He groaned and crumpled.

Throwing me a dirty look, Abu Layl said, "Why did you do that?" He struck me in the chest with his gun, pitching me into the young woman. Then he kicked the agent in the groin. When another bandit put a gun to the fallen man's temple, Abu Layl said, "No, don't kill him. Let *me* go to heaven." Militant Shiites believed that paradise awaited anyone who killed the men of Saddam.

Fixing a small bayonet to his gun, Abu Layl told his men to bind the agent's hands and hold him to the side. Nodding at me, he asked, "What's the problem with him?"

"He was laughing," one of the bandits replied.

Abu Layl lowered his nose to my mouth, pressed my cheeks with his hand, and said, "I hear you are a student. Are you drunk?"

I said, "No, I'm not drunk."

"Then why did you laugh?"

I explained about the stubborn sheikh, how he had taken my seat and paid the price. I got caught up in the story, and soon Abu Layl was laughing! He stepped toward the young woman and said, "Your husband laughed, but now you will cry." (I thought it better not to correct his error just then.) His aides opened the woman's bag and reported that she was a teacher. Then they found something else: a receipt for a donation to the Baath Party.

"Ah," said Abu Layl. "*Show* us." His meaning became clear when his aide told the woman to undress. In response, she spat on him. "*Take* her," the bandit king said. As she struggled in their arms, the outlaws cut her blouse to the waist with a bayonet.

This woman was in God's hands. I suggested to Abu Layl that I might board the bus.

"Not so fast," Abu Layl said. "Getting out of here is not as easy as getting in. I see that you have very nice teeth. We'll break them so you won't laugh anymore."

At that moment the woman broke free and clutched my arm, crying, "For God's sake, please help me!"

My heart went out to her, but I couldn't take on twenty armed men. Vexed, I asked her, "Why did you get yourself involved with the party?"

Sobbing, she said, "I'm not involved. I collected a donation from the schoolchildren and sent it in, so the receipt came to me."

"*Austad,* this is true," I said, addressing Abu Layl with respect. "I have seen this happen in my own school. When the students give their money to the party, they send it through their teacher."

Abu Layl looked at his men and said, "Didn't I tell you to break his teeth?" As the leader stalked off to tend to the security agent, his aide raised his gun toward my mouth. Knowing I could not stop him, I threw my hands over my face. They absorbed the first blow. My chest took the next one, and then my rib cage, and I felt a knifing pain. As I fell to the ground, the woman bent to help me. The aide pulled at her roughly, further tearing her blouse.

The wise path, I thought, was to let her go. If they were determined to kill this woman for being a Baathist, it would do me no good to link my fate with hers.

From just up the road came an inhuman scream: the agent. Abu Layl was having his revenge. *This is my end,* I thought. I took stock. I had not led a life of perfect virtue. I drank more often than I prayed; my eye wandered to the next pretty face. If there was a hell, I was destined to fall there. But since I was going anyway, why not leave doing good? Why not leave as a *man?*

I reached out and held the woman close. Even as the bandit pounded us with his rifle butt, I would not let her go.

"What is this?" It was Abu Layl's number two, a man they called M'shalesh, who'd been in charge of the looting. His voice was harsh and rasping, like something was broken inside his throat. And then: "Shoot them!"

"Sayyid!" I addressed him as you would a high Shiite cleric, a follower of Ali. "Sayyid, why are you doing this? We are not Baathists."

"You are not Baathists?" He seemed confused. "Then what are you?"

I had to talk fast. "I am a student, and this woman is a teacher. We are in the same boat as you. We run from Saddam, and now you come to attack us."

I paused, seeking my opening. "Sayyid," I said, "would you want your sister treated this way? Your wife?"

Before he could answer, two bandits came running with a square of metal and glass from a passenger's bag: "M'shalesh, what is this?"

I took a look and said, "It's a solar energy cell. It can be useful for you—it can run a radio, or a shaver."

M'shalesh was delighted, since his band lived without electricity. I cannot say if I pricked his conscience or he merely returned a favor. He seemed to measure us and then he said, "Okay, go to the bus." They were through here. It was time for their getaway.

As I followed the young woman up the steps, my feet

were light. I was a lucky man—I had new life! When I got on board, my friend stared as though I were a ghost.

The scene inside was grisly. I'd been wrong about the agent. The bandits had not killed him; they did not do him that favor. With their bayonets they had gouged his eyes and cut out his tongue, then tossed his writhing body into the aisle. Someone had covered his face with a towel. The old woman who'd been shot wore a newspaper shroud. The rest cowered in their seats.

Up ahead, I could see the young woman's reflection in the driver's rearview mirror. She had a sweater on to cover her shredded blouse. While others cursed or cried, she stayed silent all the way to Nasiriya. When her husband talked to her, she would not reply. Her face was blank.

As with all highway crimes, we were taken to the Governate and brought to a large hall. They gave us clothes and boots and a few coins, and took our names for future questioning. The young woman stalked about the room, sandals in hand, demanding, "Where is he? Where is he?" When she found her husband, she said loudly, "I am not going home with you. From today onward, you are not a man—I want to divorce you!" She slammed his head with her shoes until others restrained her.

I used my coins for cab fare to Saddam Hospital. It hurt me to breathe, and I wanted to check for a broken rib. After the X rays were negative, my family took me home.

The next day, as we sat for dinner, provincial security came to our house. Baba tried to show them in, asked if they had eaten. But they stayed at the door and said, "We need Mohammed."

I felt a sudden chill. There was little to fear from the mangled agent, who was in no shape to identify me. But what if someone else had glimpsed my kick from the bus?

For al-Hbeiba, already on edge, this new crisis was

intolerable. She started crying and said, "I'll go with you."

This was my first clash with my mother, and my last one. I told her, "Am I a child, that you have to go with me?" When she insisted, I slammed the door behind me and left with the two plainclothes officers.

They brought me to a reception space at the Governate, where I joined three fellow passengers: the pretty young woman, with a colorful bump on her head; the old crone in the *abaya,* who looked anxious; and a homely young man I vaguely remembered.

"Where have you been?" said the young woman, whose name was Fatat. "I looked for you yesterday and couldn't find you. My family wants to thank you, my tribe wants to thank you, for helping me. How are you doing?"

I said, "Not well, because I don't know what they want from us here."

We were led to an inner office, where a Captain Amar sat smugly behind a big desk. He acknowledged each of us by name, except for the homely one—*an informant,* I thought. I nervously struck up a conversation: "You did well—you survived."

No response. I felt myself sweating.

The captain said, "What did you people do on the bus?" To the old woman: "How about you?"

She babbled, "God keep President Saddam! Give him long life! God keep President Saddam!" She was a feeble thing, easily eighty years old.

Amar raised his glass ashtray and threw it, hard. It hit the woman in the chest; she stopped exalting Saddam. I was shocked. If this was the beginning, how would it finish?

The captain said to her, *"Now* you are singing a different tune. How could you inform on our agent on the bus?"

When the crone tried to deny it, the homely man cut her off: "Don't lie, I saw you." He turned to me and said,

"Didn't she inform?" With his large nose and small, piggy eyes, he was repellent in every way.

The captain said to me, "Save yourself before you save this old lady."

I knew the game they were playing and wanted no part of it. I said, "I was very scared—I didn't pay attention." Fatat said she was too far in the front to see what had happened. Our testimony was irrelevant, as they already had their evidence. Amar rang his bell. The old woman screeched her innocence as they took her away.

"Now," the captain said, "tell me about these robbers. Can you give me any names?" Fatat and I told him everything we knew, naming Abu Layl and M'shalesh. Still, it seemed to bother Amar that the bandits had not killed us. Might we be in league with them?

Fortunately, Fatat was an attractive woman. The captain's line of questioning soon led him to the fact that she'd already filed for divorce. "You deserve to be divorced," he said. "Your husband is a coward. If I had a woman like you . . ."

Fatat laughed coquettishly. They exchanged phone numbers and agreed to meet the next night. Then she was gone, leaving all eyes on me.

"I'm a student," I explained, when asked what I did in Nasiriya, "and I help with my family's stores."

Abu Layl was suddenly forgotten. When the homely man heard that we owned a perfume shop, he got excited. "Do you have Paris?" he said. "And Cobra?"

"Yes, we do," I replied. "We have the real ones, from France and India, and we also have the Iraqi imitations."

He said, "I want the real ones." The interrogation was over. The captain and his informant drove me to the perfume shop, took what they pleased, and said good night.

In Iraq there were many ways to be robbed, not all of them at gunpoint.

CHAPTER NINE

MY MOTHER never worked outside the home or traveled alone outside the city. Though she wore her hair in the modern style, short and parted to one side, she was a traditional Arab wife, dependent upon Baba for everything. They rarely argued. She deferred to his judgment and probably feared him a little bit, too. She could not bear for bad feelings to come between them.

Al-Hbeiba wanted more than anything to have a doctor in the family. When my sister Zayneb fell in love with a young man out of medical school, she was thrilled. Baba, though, had his doubts. The young man came from a modest family; he needed help to establish his practice. Did he want Zayneb or her dowry? So Baba told him, "We are happy to give you our daughter, but you will take her with her clothing and jewelry and nothing else. You will get nothing from what I own."

The fiancé said that all he cared about was Zayneb. He would bring his family the following Thursday to make the engagement official.

When my mother heard what had happened, she was

aghast: "This is not the way you treat someone who asks for your daughter's hand! If you test people like that, they will run away. Zayneb will never get married."

Her fears were realized the next Thursday, when the fiancé failed to show up. The engagement was off, and my mother was so upset that she could not look at Baba. In truth, however, my father had been proven right. The young man was a blatant profiteer. Had he been true to Zayneb, Baba would gladly have helped him with his practice.

Zayneb wound up marrying a man with a car repair shop—a solid living, if low in status. Not until Hassan married Hamida would we get a doctor in the family.

But al-Hbeiba would never see it.

I was twenty-four, at school in Basra, when my mother—still young and lovely—was diagnosed with kidney failure. Over the next two months, I came home so often that my studies suffered. But what could I do? When I was away, I thought only of her. I felt a terrible weight drawing me into darkness.

One day she told me, "If you want to help me, go back to college. When I see you heading back with your books, I'll feel better. I don't want people to say you did not succeed in your schooling and career. I want you to be successful."

"But I need to be with you."

Al-Hbeiba said, "Look, call me when you arrive in Basra. If I am doing badly, I'll let you know, and you can come back."

Reluctantly, I agreed.

"And, Mohammed? Don't think that if I die, I won't be with you."

I was crying when I bent to kiss her good-bye. To make me smile, she told me, "I'm giving you this kiss, but you give it to your girlfriend."

Those were her last words to me.

When I called home from Basra, they told me she was sleeping and could not be awakened. The next day, no one answered the phone. That afternoon, my uncle Nouri made the three-hour drive from Nasiriya to my dormitory. He asked me to pack for a short vacation.

I said, "But I have class tomorrow."

Nouri said, "The class has been canceled."

I said, "Just tell me—what's wrong with al-Hbeiba?"

My uncle started crying, and I knew. Two hours after I'd left, she went into the hospital and died there. As our religion prescribed, she was buried the same day. I could not have made it back in time, and my family knew that I had a test at school. They spared me the news until Nouri could see me in person.

By the time I made it home, mourners had gathered from the neighboring villages. I accepted their condolences without hearing. I went to the gravesite and asked my mother's forgiveness for not being with her at the end.

I missed al-Hbeiba's affection, her small touches around our home. I could not bring myself to enter her room and see her knickknacks and jewelry. When we sat at the dining room table, her chair was empty—I could not eat. With my brothers in agreement, I told my father that we must sell the house. It was too sad to stay there.

But Baba said, "If you loved your mother, this is where you can remember her. Even if I lost everything else, I would not lose this house."

My father was a sheikh and the son of a sheikh. He knew about families and the land, and how they endured together.

Our least popular professor at Basra was an ex–security officer named Naqib, who taught a second-year course on the penal code. I arrived early at his class one Wednesday afternoon with suitcase in hand, as usual. Like the other boarding students, I looked forward to my long weekends, when I might see Maha or replenish my food supplies and pocket money at home.

I knew something was up when none of my classmates had their luggage. One of them said, "Haven't you heard? We can't go home this weekend—the teacher has scheduled a big test for Saturday."

I was beside myself. "Oh my God," I ranted. "We Iraqis never make objections. What would happen if we did it today? What do we have to lose?"

"You are right, Mohammed—you tell him!"

It was easy to stir these students up, but I knew that none of them would challenge Naqib. We were too conditioned, too afraid.

Our teacher outdid himself that day. No sooner had we taken our seats than he announced an oral pop quiz. Everyone groaned. I sat behind a woman named Janelle, whose hair was styled in a halo of curls. I said, "I love the way you've done your hair. I'll pay you to keep it that way so I can hide behind you."

When Naqib asked Janelle why she was laughing, she blushed and looked at me. The teacher told me, "You are a lamb!"—a great insult, meaning I was weak.

Already peeved about our lost weekend, I answered, "If I am a lamb, you are my father!"

As the class rocked with laughter, Naqib slammed his desk and said, "You call me a *sheep?* I'll take care of you— you have sentenced yourself to failure! We are going to the dean's office, now!"

The teacher told the dean his side of the story and demanded that I be expelled. Then I said, "He told me that I am a lamb, and all I said was that he is our father. Is there anything wrong with that? Our teacher *is* our father, in a sense."

The dean couldn't stop himself from smiling, and suggested that I wait outside his office. After a long talk with Naqib, he told me to go back to my classroom. I could see our teacher boiling whenever he looked my way.

To survive the boredom of law school, I took theater classes on the side. Thanks to my kung fu ability, I had won the starring role for National Army Day. Our big production was set for that Saturday night. Every big wheel in the province would be there, including Governor Tareq Tikriti.

Having stayed in Basra to study for the test, I was ironing my costume in my room for Friday's dress rehearsal. Just before I was to leave, the dormitory superintendent opened my door and three strangers filed in, each one bigger than the last. The one in the lead, a security captain, was blond and sarcastic. He said, "Are you *Professor* Mohammed?"

"Yes, I am Mohammed."

"Come along, we have business with you."

I said, "I don't even know you."

The captain opened his jacket to show me his gun and a pair of handcuffs. "Let me introduce myself with these," he said, cuffing my wrists.

I'd already guessed what had happened. One of Naqib's students had informed about my sedition before the class. It hadn't taken long—even on a weekend—to get these three to pay me a visit.

I told the captain, "I didn't do anything."

The captain said, "I hear you have a long tongue. We'll make it shorter for you."

I asked him to wait a minute—to let me call Tareq Tikriti.

"What is your relationship with the governor?"

I said, "He's responsible for the Army Day performance, and I am the main character."

"Stay here," the captain said. "*I need to make a phone call.*" He returned with a friendlier tone: "We'll just go for a short visit. The boss wants to talk to you."

The black Subaru took half an hour to reach a large building in downtown Basra. They led me down a long corridor, to an unmarked door. The captain rapped on it and said, "Sir, here's Mohammed Odeh."

"Come in—I want to see this one. So *you* are the big rabble-rouser? You're a little shrimp! I expected a huge man to be making all this trouble." The boss introduced himself as Staff Major Sabbar. He was a smart-looking, gray-haired man, about fifty and very trim—certainly no one to be trifled with. But you must understand that I was a rash young man in those days. I had an uncle high in the Baath Party, and a governor with a front-row seat to watch me perform in twenty-four hours. I felt untouchable. Riled by the remark about my height, I replied with my favorite saying from the David-and-Goliath story: "The stone you don't respect can hurt you."

The slap hit me full in the face. I was used to stronger blows in the ring; I did not fall.

Sabbar said, "Don't you know that your little student protest was mutiny against the government? Do you want to spend the rest of your life in jail? Do you want to die?"

I said, "I did not do anything."

He slapped me again and said, "Now listen to me well. I will teach you how to answer—not by words, but by hand. Get me Abdulla the Gorilla!"

The heavy feet stamped down the corridor and into the room like some movie monster's. I found myself staring at Abdulla's belly button. He was dark and hideous. His two front teeth were jagged and broken. An old knife wound stretched from his temple to the corner of his mouth.

Abdulla was a polite gorilla. He extended his hand. My own vanished inside it.

"Hi," I said. "How are you?" The giant said nothing. He was waiting for his cue.

Sabbar said to Abdulla, "Our little friend doesn't seem to understand me so well. So I will talk to you, and you'll explain to him. Now—beat him up!"

The captain said, "Not here, sir. The carpet will be full of blood."

The major pondered this for a moment and said, "All right, I will try one more time. Saddam over all!"

I nodded and said, "Certainly."

"The Baath Party is one and all!"

"Of course."

"When someone has a long tongue, we cut it shorter."

"Yes," I said, "I understand."

"And when someone writes something we don't like, we cut his leg."

I could not help myself. "Yes, sir," I said. "We cut his *hand.*"

The other officers laughed at my correction. But Abdulla did not laugh, and neither did Sabbar. "Go!" he told the gorilla. "Teach him a lesson."

Instinctively, I crouched into a defensive stance. Abdulla drummed on his barrel chest with his fists. Then he lumbered my way and unloaded a kick at my stomach. It was well aimed but sluggish, and I dodged it easily.

"Stop!" the major said. "What is it that you play?" When I told him of my black belt, and my starring role for National Army Day, he was taken aback. He shot the captain a dark look and said, "Why didn't someone tell me about this man?"

It seemed that the captain and the major's secretary had gotten their wires crossed. Sabbar was a martial-arts enthusiast himself; he had tickets for Saturday night. "All right," he said. "I could tell you were good the first time I slapped you and you did not fall. I like a strong man. Get dressed and get out of here."

Truth to tell, I appreciated this Sabbar. He had dealt with me decently. I signed a pledge to desist from further protests. My student organizing days were over.

My romance with Maha lasted into my Basra years. We traded visits, and distance was the only thing between us. In Baghdad we'd spend hours at a kabob restaurant called al-Multaqa, making each other laugh. Or we'd stroll through groves of eucalyptus, content to say nothing at all.

My parents worried about our age difference. They did not want us to marry, but I knew they would come around. I was sure that Maha and I would spend our lives together.

As an Iraqi, I should have known that there were no certainties in life.

We had our last phone call almost three years to the day after she'd surprised me at the farm. I told her that I loved her and would see her in Baghdad at the end of the week. I remember that we hung up at exactly 8 P.M.

The next morning, her family found her dead in her bedroom. The doctors said it was heart failure, one of those strange short circuits that could strike anyone out of the sky.

I went to the hospital unable to accept that Maha was dead. I went to the grave and could not believe it. Even now, I can see her laughing and alive.

Like my uncle Ahyal, the faithful suitor, I swore that I would never marry.

After two years I transferred to Baghdad University, Basra's private, more prestigious sister school. The law faculty was packed with influential judges and Justice Ministry officials. For ambitious young Iraqis, this was the place to get ahead.

Saddam's face was everywhere, even on the campus computer screensavers. (As an alternative, you could choose either Uday or Qusay, the president's sons.) Students avoided any hint of politics outside the party. There was a famous "Freedom Wall" in the administration building, a large white marker board. A Baathist student leader told us, "If you have anything to suggest, put it on the wall, and I will take care of it."

No one trusted him, of course, and it didn't help that campus security was based across the street. The wall stayed white all year, with one disastrous exception. It began when my friends and I cut our criminal law class to go to a daytime party at another local college. When our teacher arrived and saw only a few women at their desks, he canceled the class. Out of spite, he locked the door—a big problem for us, since we customarily left our dinars inside the books on our seats. Now we had no way of getting at our meal money. We were fighting mad.

We had a classmate named Hadeer who'd been released from a mental asylum the year before. She knew no fear. If you challenged her to do something, she would do it without hesitation. I told her, "What this teacher did to us is not right. This is oppression, to lock us out from our money. We need to put this on the Freedom Wall!"

After my brush with security in Basra, I was hedging my bets. If the administration accepted our protest for review, my friends and I would come forward. If it backfired, we assumed that they'd go easy on Hadeer because of her history. We told her what to write:

We submit this complaint to the University Board, against the instructor Abdul Jalil, who is oppressing us and has violated our freedom. This is not appropriate treatment for college students.

The protest backfired. The board denied a review and hunted down the culprit. Though Hadeer had not signed her name, they traced her though her handwriting. The next week they sent her back to the asylum.

As the regime saw it, you had to be either treasonous or insane to protest in Iraq.

I felt sick with guilt. But there was nothing we could do, or so we told ourselves.

Of the thirty-five hundred students at Baghdad University, only ninety-five were enrolled at the law college, just twenty-one in my class. We were a select group, ripe for enlistment into the regime's vast web of informants. As I later found out, half of my twenty classmates were on the party's payroll. (Even the broom closets had ears. Our janitor was a captain for campus security.) In addition to a stipend and a waiver of fees, the party offered a promise of good grades to any who would watch and tell. Few had the backbone to turn this package down. As a result, we spoke cautiously except among our closest friends.

Domoua was a law student with a dancer's body and dark pools for eyes. We were friends, though I found her conceited. I was on campus one day when I saw her car hit from behind. As they checked the damage, the other driver became vulgar with her.

I went to the scene with two other classmates, Nabil and Samir. When we scolded the man for mistreating a woman, he waved a revolver at us. He was an officer in the Bureau of Military Industrialization, and he liked to throw his weight around. We told him to put the gun away—we'd be happy to fight him if that's what he wanted. He threatened to call security.

Nabil said, "Go ahead and call."

As the man pulled out his phone, I told Domoua, "If you don't have enough money to bribe security, let's make a deal with this guy."

And she said, "I have my checkbook, but I want to see what security will do."

Ten minutes later, seven men from general security arrived in two cars. Their chief said, "Where are the people making all this trouble? We are here for them."

I was terrified. Samir was terrified. Nabil and Domoua looked as calm as could be. Then Nabil whispered something to the chief, who said, "Okay, you can go."

Now I got it—Nabil was one of *them.* I grabbed his sleeve and said, "Can you help Domoua?"

He looked at me coldly and said, "No, it's everyone for themselves. I'm not getting involved." I was miffed. I thought Nabil was being selfish.

Samir had slipped off, leaving me and Domoua to handle the officer. Pointing at me, he said, "That man was rude." The chief nodded, and two of his men took my arms to arrest me. With more students gathering around, it was an embarrassing spectacle.

"Release him," Domoua said. "This is my problem. Mohammed has nothing to do with it. Let him go."

The chief leered at her. "I see, this must be your boyfriend. That's very cozy, isn't it?"

Domoua was hot-blooded. Staring daggers at the chief, she said, "Now I will tell you who I am: Number seventeen slash one thousand of the Fourth Division."

The chief said, "Who is your superior?"

Domoua said, "We are from Ghaleb's unit, education branch." I was astonished—Domoua was in security, too! Not only that, but she outranked these men from the general office, at least while they were on campus.

And the chief said, "Why didn't you two identify yourselves before, like your friend?" He was sorry. The military officer was *very* sorry.

Feeling cocky, I said, "This isn't the end of this. I want

the name of this officer, and who is his boss? And I want those men who held me to apologize." If I was supposed to be an agent, I might as well play the part.

Prodded by their chief, the two underlings kissed my cheeks, our way of extreme apology, and said, "We're sorry, sir, we didn't know."

"That's *better*," I said.

After they had left, I turned to Domoua and said, "Are you on the level? Should we come to class tomorrow, or should we run away?"

And Domoua said, "Don't worry, you can come to class." As a favor, she asked me not to tell anyone about her side job. She'd revealed herself only because she'd lost her temper with the chief.

By the next day, the entire law school knew about her affiliation—and mine, or so they thought. When I walked into the student union, my classmates cut off their conversation and started praising the state, the Baath Party, Saddam, Chemical Ali—it was disgusting. They treated me that way till the exam period, when a professor refused to adjust a test date and allow me to attend a martial arts tournament. Then they *knew* I wasn't in security, and all returned to normal.

After doing well in some local tournaments, I founded a campus kung fu club. I was up to fifty members when two men barged into my office without knocking. They needed private lessons, they said—and did I teach a move called the Destroyer? Within a month's time, they explained, they had to be able to kill a security officer— "that bastard!"—named Saad.

On my guard, I said that I taught a version of the Destroyer for use against unarmed muggers, not cops. I could not help them with this Saad, I said. It was not my job to assist in a murder.

The men dropped their deception. One was from security, the other intelligence. They asked about my club members, especially the men. Were any from families who had been in trouble with the Baath Party? With relatives who'd been jailed or executed?

I said, "I don't know anything about that. I train their bodies, not their minds."

Maher, the one from security, said, "We need you to train their bodies *and* their minds." He pushed me to find out more about my students' political views. In return, I would get the standard compensation.

Knowing it would be dangerous to turn him down, I agreed. But when we met a week later, I had nothing for him—I had tried but failed.

Maher was frustrated: "You are not qualified to work with us." He thought I was incompetent. It never occurred to him that I might have a conscience.

CHAPTER TEN

IN THE SPRING OF 1995, with my graduation approaching, Baba and my brothers and I ordered identical, tailored cotton suits—dark green and very stylish, we thought. Because I was an honor student, Baba would be escorting me to the stage for my diploma. I was proud of my father and how distinguished he looked.

A few hours before the ceremony, as we were leaving for our drive from Nasiriya to Basra, a friend of Baba's dropped by. He looked at our suits and said, "What is this? You look like members of a band!" As my father winced, the busybody added, "You are a prominent man, a sheikh. You should dress like a sheikh!"

The men of my father's generation moved in two worlds. They wore Western clothes in town and traditional dress in the countryside. Baba was uncertain—what should he do? Deaf to my pleadings, he stripped off his suit and changed into his black silk *juppa,* which draped to his ankles. He added his fanciest kaffiyeh, with its loose white scarf and black headband.

He came out of his room and said to his children, "How

do I look?" This was a ritual in our family; we always assured him that he looked splendid.

But this time I said, "No, it is no good. I don't like your outfit—you should change it."

Baba glowered at me. "Am I your father or are you my father? You need to be more respectful."

I said, "But you always taught me to be honest."

And Baba said, "Yes, be honest, but this is not the right time."

Stung by his rebuke, I said little during the drive to Basra. When we entered the campus auditorium, where all the parents had turned out in their finest European styles, I felt everyone staring at us. When my name was called and I rose with Baba to go to the stage, I half expected someone to burst into giggles.

As we shook hands with the dean, he told us, "I would like to see you afterward in my office." *What now?* I wondered.

The dean said to Baba, "Please don't be embarrassed, but I loved what you wore tonight. Can you tell me where you bought these clothes? I'd like a set for receptions and special occasions."

In the Arab tradition, of course, my father gave the dean his *juppa* as a gift.

Once again I felt embarrassed, but not for Baba this time. He was true to himself, unafraid of what others might think. Though I had my diploma, I still had much to learn.

Ali Khayoun was a handsome man with a strong personality. He looked much like my mother: round face, small mouth, and expressive, almond-shaped eyes. But where al-Hbeiba was gentle and sentimental, her oldest brother was tough and cunning, a stickler for precision. He was professionally discreet, with a voice so soft you could hardly hear him on the phone. He commanded your attention without shouting.

When I was a boy, Ali worked as a junior officer in the Ministry of Defense. Still a bachelor, he often took our family to dinner or on outings to a park or playground. On weekends he'd come down to al-Abthar for the duck hunting.

Everything changed when Ali married a younger woman from Saddam's tribe in the north, in the area near Tikrit. My uncle took the unusual step of adding a new name from his wife's clan—he was now Ali Khayoun al-Nasiri. To gain the regime's trust, he even became a Sunni.

Everything changed. Soon my uncle was secretary to the defense minister, Adnan Tulfah—Saddam's cousin, brother-in-law, and closest boyhood friend. (After a family feud provoked by Saddam's mistress, Tulfah died in 1989 in a strange helicopter crash.) By the time I made it to law school, Ali was a big man in the Baath Party, with two more jobs: director of political guidance for the Ministry of Defense and personal adviser to Qusay Hussein, the second most powerful man in Iraq. Unlike older brother Uday, a hopeless hothead, Qusay was a cold-blooded shark, the regime's version of Michael Corleone. He ran all internal security forces. Everyone feared him.

With Ali's new status came the high life. He had three houses in the best neighborhoods in Baghdad: one for his wife and two daughters, one for his in-laws, one for his mistresses. His family home was a sprawling villa with modern Italian furniture and glass everywhere. He had a big-screen television, three Mercedes with drivers, and seven bodyguards.

He never forgot where his wealth came from. Portraits of the president and his sons crowded the desk in my uncle's home office. Framed photographs lined the wall: Ali next to Saddam, or Qusay, or Saddam and Qusay. The face of his watch bore a picture of Saddam. So did his gold pen. When he went on about Saddam's latest stroke of genius, Baba would say, "Stop it now. We want to hear your news, not Saddam's news."

And Ali would say, "My news *is* Saddam's news." He was a true believer.

My uncle became a very busy man. When we dropped by his office, he'd come out to say hello, then rush on to his next meeting. He rarely visited the farm anymore; he needed clearance to go to other provinces. He was important enough to have enemies.

Baba thought it just as well that we saw less of him. There were vigilant anti-regime elements in Nasiriya, and my father had no desire to publicize our relationship. On the other hand, it was good to know that Ali was well connected, just in case.

As I got older, I realized that my uncle was a selfish man. It was "Saddam over all!" and Qusay next, but Ali wasn't shy about taking his share, either. Once Baba heard about some good Italian wood for sale by the government; Saddam had decided the color was wrong for the interior of his new palace. After my father made an offer on the lot, he asked Ali to put in a word for him. My uncle thought about it, put in his own bid, and made a fat profit.

Still, I counted on Ali's help in landing a job after graduation. For years he had urged me to join the Baath Party, and I'd say, "You are already there with Saddam, there's no room for me." I tried to be diplomatic, but Ali could tell that I had no interest. While I knew hundreds of lukewarm members who had joined out of convenience, I hated the idea of doing the party's dirty work. For example, suppose that one of my old Nasiriya friends went AWOL from the army. My duty as a Baathist would be to track him down and bring him in. Where was the honor in that?

Looking back, I was arrogant about my family connections, my big-shot uncle. I remembered Ali once saying that when I applied for membership, he could fix my papers to show that I'd been in forever. I thought I could put off my decision to the very end.

In school I majored in international law, which put me in line for a post with the Foreign Ministry. I was eager for the job, a shortcut to fulfilling al-Hbeiba's dream for me. In the ministry a clever man could land a judgeship in three years, instead of the usual ten.

At my interview, my qualifications were in good order: scholastic record, reputation, appearance. The interviewer said, "I don't think there will be any problem here. Of course, we will need your party membership card."

I no longer had a choice. I went to my local Baath official, who said, "Oh, now that you need us, you want to be a member? You're too frivolous for the party."

I stormed out and went to my uncle. He looked at me with exasperation: "Of course, party members get preference—you needed to join *before* you applied." I had waited too long; there was nothing he could do. Then he said, "You didn't use my name at the Foreign Ministry, did you? It wouldn't look good for me to have a nephew outside the party."

I was hurt and disappointed, but also relieved; I'd been ambivalent about taking the party's path to success. I opened a practice in marital law, with a specialty in divorce cases. My clientele was largely female, since men needed little help in these matters. For the most part, an Iraqi husband could be done with his wife (and off to another) simply by saying so and leaving. For the unhappy wife, the law was more complicated. While it wasn't so hard to get the divorce, she'd have to wait seven years to remarry—unless she could show that she'd ended the marriage with good cause. A poor-earning husband was solid grounds for splitting up. A wife-beating husband was not, unless bones had been broken. If a man was a deserter or in jail, that was open-and-shut; I drew a steady income from such cases. But other situations made me earn my fee.

One woman came into my office crying, in itself not unusual. She was a middle-aged bank employee who had lost her first husband in the Iran-Iraq War. As compensation, she received a car, funds for a new house, and a monthly income. She also drew the attention of an Egyptian, one of many who came to Iraq to work during the war. The Egyptian drank a lot, but life was tolerable until the woman's daughter by her first marriage turned seventeen.

"When I go to work at the bank," the woman sniffled, "he flirts with my Ilham. He attacks her. He molests her." The woman could not discuss the problem publicly, much less go the police, because Ilham would be disgraced and unmarriageable. Their male relatives might even kill the girl to preserve the family's honor. The only solution was a quiet divorce, which was where I came in.

Ilham came in with her mother the next day for a follow-up, and she put me off from the start. Her shoulder-length, light brown hair was streaked with blond highlights. She wore layers of makeup—thick mascara, fake eyelashes. Her lipstick matched her brown high-heeled shoes. I was not a holy man when it came to such temptations, but this girl looked cheap. She seemed strangely cool and composed for an abuse victim. I became skeptical.

Taking a microcassette recorder from my desk, I told

Ilham, "After your mother goes to work and your stepfather comes to see you, don't avoid him. Let him talk, but secretly turn this recorder on. Then we will have some evidence."

As we began to discuss my fee, the mother said, "I have to go fix my makeup. Why don't you talk to Ilham till I get back?"

No sooner had the mother left than Ilham shifted in her chair and crossed her legs. Her *abaya* parted far enough to reveal what looked like a miniskirt underneath, hiked up too high for modesty. She said to me, "Forget the case for a moment. All life should not be work. Besides, aren't you satisfied with the dinars you will get from this?"

Her brashness angered me. No one spoke of money in such a familiar tone at a first meeting. But Ilham was just getting started. She stood and opened her *abaya*. As she leaned over my desk, I could see that it wasn't a miniskirt, after all—it was a sheer blue negligee, low cut, set off by the girl's fair skin.

"How dare you speak to me about money!" I sputtered. "How dare you talk to me that way!"

Ilham leaned over even farther and said, "Because of *this*." She dipped her shoulder forward. As they say in Iraq, she was very well qualified.

I felt weak in the knees, and not from lust. My secretary, a sharp-tongued woman who didn't mind a little gossip, could happen in at any moment. Baba might drop by, or one of my brothers. My reputation hung by a thread. I said, "Don't you want to sit back down? Your mother won't be happy with you when she gets back."

Ilham said, "Don't worry, she's not coming back. Besides, a big lawyer like you should not be scared. How are you going to win our case if you're scared?" She nudged the back of my swivel chair, which rocked alarmingly. I braced my leg to stop it and got to my feet. This girl was fishing in dirty water. I opened the door and told her good-bye.

Ilham placed her hand on my chest and said, "Where are you going?" She nuzzled closer to me. I worried that her makeup would rub off on my white shirt.

I said, "*You* have to go—and I'll be calling your mother about this!"

Pushing away from me, the girl said, "You have no feelings!" She left the office spitting curses.

When I got through to the mother, I said, "As you fed your daughter food, you should also have fed her some manners." She apologized, but I did not believe her. I was convinced that she had left Ilham to see if I might take my fee in kind.

I should have dropped the case then and there, but those were hard times in Iraq. A divorce lawyer couldn't be too choosy about his clients. I would get a fair down payment, and a lot more if I won in court.

The next day Ilham returned, dressed in slacks and a proper blouse. She brought my down payment and—happy surprise—the tape recorder. I thought, *This case might not be so hard, after all.* I pressed the play button and heard the sound of kissing, then Ilham's voice: "Aren't you married to my mother? Don't you love my mother? Why are you doing this to me?" She didn't seem to be fighting the stepfather tooth and nail, but the tape was useful from a legal point of view. The Egyptian said something about getting "a sexy movie" for them to watch—and then the tape ran out. Ilham had recorded some music before getting down to business.

I filed preliminary motions to get the case moving. Two days later, the mother came and asked for the tape. No explanation—she just wanted me to drop the case. I didn't give her more thought until six months later, when I saw mother and daughter, draped in black, being ushered into the courthouse by a policeman I knew. When they spotted me, they covered their faces.

Later that day I ran into the officer. "What's up with the two women?"

"Do you know them?"

"Not really, just curious."

The officer said, "It's a sad case. These thieves came into their house and killed the husband, and now they have come to give testimony." According to the police account, the family was asleep on a rainy night. The thieves broke in a door, and Ilham's mother asked her husband to check and see—she thought it was the wind. When he went downstairs, he was stabbed three times, once in the chest.

They never caught the thieves.

CHAPTER ELEVEN

AFTER BABA CAME DOWN with a kidney ailment, I dropped him at Saddam Hospital for some tests. When I returned three hours later, he told me that his sonogram had yet to be done: "There is a nurse named Iman, she did not show up."

My father felt poorly that day, and I had no patience for someone who could not be bothered to come to work. I stomped down the corridor and found two nurses chatting by their station. I went up to them and said, "Where is the hospital director?"

"Why are you so upset?" said the one who was taking off her coat. "Calm down, let me help you." She was a pretty woman, with curly hair and a great smile, but I wasn't noticing.

"*You* calm down," I said. "How can I calm down when this Princess Iman is three hours late to take care of my father?"

The nurse said, "Why do you need the director?"

"I want to complain so that next time this Iman won't be late."

She steered me to the ground floor, but the director was in a meeting. When I went back up to see Baba, I found the nurse examining him. I thanked her for her kindness, and said I would mention her when I made my complaint. She laughed; she had a nice laugh.

When I managed to buttonhole the director, I told him, "You have a very good hospital here, but little things can hurt your reputation."

"What's wrong?" he said.

"The nurse Iman never showed up today. People are sick here, and nurses should be on time."

The director said, "I agree with you, but Iman is here today."

"No, I was just up on the fourth floor, and she wasn't there." As I railed on, the director asked his secretary to find Iman. "You know," I said, "you should punish her."

"No, I am not going to punish her."

"Then I will go over your head." My temper was getting the better of me. "What kind of hospital is this where the nurses don't show up for—"

"Yes, Doctor?" It was the pretty nurse who had seen my father.

The director said, "This is the Iman who did not show up today."

Perplexed, I said, "*This* is not Iman. This is the nurse who helped me!"

"No," the director said, "this *is* Iman."

I was mortified; I actually blushed. Iman was indignant. Once Baba was taken care of, she hadn't thought I was actually going to complain. Yes, she said, she was late, but she had a good reason and she'd called in ahead of time.

After I thanked the director and turned to leave, he called me back to thank Iman and apologize. My pride bruised, I said, "I'll decide if I should apologize or not."

And Iman said, "Whether he thanks me or not, it doesn't matter."

So cheeky! I was full of myself because the Iraqi minister of health had bought a house in Baghdad from Baba. I told her, "Okay, so you are Iman and I am Mohammed. If I don't get you transferred somewhere very far away, then I won't be Mohammed! You will spend your whole salary on buses and cabs!"

Iman looked at me coolly and said, "Go right ahead."

I was so flustered that I took Baba home without his medical papers. To relax, I went to work in our garden. When Hassan called back to me that some lady had brought Baba's papers, I told him to drive her home.

Then I heard a scream and dropped everything to run out front. Hassan had left the door open and our pony-size dog was attacking Iman, still in her nurse's whites. I pulled the dog off. Iman was shaken but unhurt.

By then the whole family had come to see about the ruckus. They brought Iman into the house and gave her juice and salt water, our remedy for fright. Soon they were all talking, and I could see that everyone liked her. Iman was gracious and down-to-earth. I felt foolish about my tantrum in the hospital.

As she left, Iman needled me, "Don't just think to complain next time, but remember to take your papers, too."

I got mad all over again. "You just brought them because you were scared that I would get you transferred."

My family was appalled. Batul said, "This is a lovely girl. She shouldn't be treated that way."

As Baba recovered, I kept bringing him to the hospital for weekly checkups. One time we had to stay late to wait for a doctor. My father got hungry, and the patient menu was too depressing to consider. I called home to see if someone could bring something, but no one picked up.

Noting our misery, Iman went to her house and came back with home-cooked dolma, grape leaves stuffed with lamb, rice, raisins, almonds, pistachios. They were exceptional dolma. I thanked her and finally apologized.

That was a turning point. Iman had struck me with her courage since the day I tried to bully her. Since then I had seen other fine qualities—her gentleness with Baba, her compassion with an ailing patient off her unit. She was a good person, this cheeky nurse.

No, she was a *great* person.

As we talked, I learned more about her family. She was the oldest of five sisters and had two brothers as well, one of whom owned a pair of movie theaters in town. Her father, called Hajji Juma, ran a delivery company. Like Baba, he had been to Mecca; like Baba, he was no extremist. Iman had lived in Nasiriya all her life. We knew the same streets and parks and restaurants.

One month after we met, we professed our love over dinner. That set the engagement process in motion. First the women of Iman's family received the women of mine: three great-aunts, my three sisters, and two of my sisters' friends, all led by Batul, the eldest, who stood in for al-Hbeiba. They confirmed my interest in wedding Iman and established that the match was compatible.

Then the men of my family went to the men of Iman's to ask for her hand and settle the dowry. To make a better show and allow Hajji Juma his rightful boasting, we gathered enough men for five cars—not only Baba and my brothers, but distant cousins from the clan.

Only two months after that did I meet Iman's parents. I felt at ease with her father, a respectful and educated man. We seemed to see eye to eye, which reassured him that Iman and I would get along well in life.

We were married on a Thursday, the traditional wedding day for Muslims: December 10, 1995. First there was a civil ceremony in court, as required by Saddam's secular state; then a ceremony with a sheikh, in the Islamic way; then a party at a downtown hotel, where we danced the *shobby* to a live band. According to the folklore of our country, if either the bride or groom treads on the other's shoe, that one will lead in their marriage. As I kissed Iman on her forehead at the hotel, I felt her foot on mine. For a moment I got upset—not that I worried about being henpecked, but because she'd scuffed my fine Italian loafer.

The next morning we went to Baba's house for the ritual slaughter of a lamb, with the meat going to the needy. As he made the sacrifice, the *boucher* recited, "By the name of God, God is great." Those words would be the foundation for our happy married life.

En route to our honeymoon in the capital, we rented a cabin on the Tigris for a day on the beach. I had a good camera on a sunny afternoon, and we took some wonderful pictures—keepsakes for our children, and their children. But there was a problem at our hotel. The Baghdad Sheraton was in a high-security zone, near an air base and Uday Hussein's palace. The state security officer insisted on taking my camera, to be returned when we left. Since every good hotel in town was booked for a festival that weekend, we had to play by their rules.

Three days later, when we checked out, the camera was missing. The officer who took it was on vacation. I told his replacement, "I don't care; I want the camera."

The security man said, "No problem. I'll take you out to our office and you can wait there until they find it." Which meant until the rooster laid eggs and the ox gave milk.

I said, "Oh, so you are changing my reservations from the Sheraton to the security office?"

The man said, "You are mocking us?" He looked at my ID. "And beyond that, you are from Nasiriya?" Which meant that I was Shiite, and therefore an enemy of the state.

"Yes, I'm from Nasiriya. Is there anything more that you want to know?"

By this point Iman was pulling me away: "Let's go, Mohammed. Forget it."

This time, though, I knew I had the upper hand. We'd been invited that night to dine with my uncle Ali Khayoun. When I told him what had happened, he promptly called the national director of security, Taha Sheikh. The director told my uncle, "By the time you drink your tea, you will have this camera."

In fact, it arrived before the tea, in less than half an hour—along with five special agents, twelve extra rolls of film, and a pair of handsome watches as wedding gifts for me and Iman.

There were times, I thought, when it was good to know people in high places.

When Iman and I married, we thought we would wait five years before having children. Like many newlyweds, we wanted to travel and enjoy some freedom before we settled down. But I also had other reasons to wait. It was the time of the sanctions, of "oil for food," and no one was happy in Iraq. The rich were apprehensive, and the poor were wretched. Even the powerful were uneasy. Not long before my wedding, the governor of Nasiriya alerted Baghdad to an insurrection in his province. When it turned out to be a false alarm, just a feud between two tribes, Saddam was incensed. He had the governor executed.

No, it was not the best time to bring new life into our

world. On the other hand, Baba was getting older and craved more grandchildren. I wanted to be a dutiful son.

When they called me from Nuns' Hospital in Baghdad on May 6, 1997, I was in court in Nasiriya. I rushed through my summation, talking even faster than usual, then gunned my car to seventy miles per hour on the highway. After flagging me down, the policeman was unmoved by the great event and wrote me a ticket. By the time I arrived, arms full of red and yellow roses, my daughter had beaten me. (We took her name from that bouquet: *Abir,* or "Scent of Flowers.") Five nurses brought her from the nursery, and I joyfully paid each one, as is the custom.

In the instant that I first saw my firstborn child, my life changed. I felt older, more responsible. I felt like a father.

Abir was a healthy infant. When I picked her up, she felt substantial. When we took her home, people on the block were surprised to hear she was a newborn. She had a full head of black hair and a serious way about her, as if she were thinking deep thoughts.

I could already predict a great future for her.

CHAPTER TWELVE

EVEN WHEN our electricity was working, Iraq was blacked out in more subtle ways. Saddam understood that to keep people down, physical force was not enough. He needed to control their minds as well. There was nonstop propaganda over the state airwaves, and next to nothing to challenge it. Telephone landlines were easily monitored, while cell phones were limited to calls within Iraq. Possession of an international satellite phone was proof of treason. When a friend of mine was found with one in his pocket, he was killed.

There were two television channels in Iraq. Shabab, or "Youth," was best known for its soap operas. Al-Joumhouria, or "Republic," aired the filtered news. The first twenty minutes of each hour were reserved for the latest glories of Saddam and his sons. Many timed their viewing to skip that segment, which seemed like one eternal rerun.

A Kurdish friend named Serwan, a high school classmate of mine, had a satellite dish smuggled in from Turkey for his fiancée. When their relationship soured, he got the dish back and sold it to me for $500. We installed it one night on my roof and hid it under a tarp.

Suddenly I had a new world, as close as my living room. For a time I sat glued to the news from Kuwait or Saudi Arabia, countries that didn't bother censoring much about Iraq. Or I'd watch al-Jazeera, the famous Arab station based in Qatar, or the BBC when I could get it. It was a funny sensation. Even as I was locked in my house, careful to keep the volume low, I felt like I was breathing fresher air.

As the novelty passed, I ranged around the dial, to action movies and all sorts of sports. I followed tennis and track and field, and most of all I loved soccer—I would always root all out for Maradona. Iman preferred horror movies and any program with disco dancing.

When Baba sat down with us after dinner one night, he noticed that we had forty channels too many. "This is very dangerous," he said. "Is it hidden safely?" It was, I said.

"Does anyone else know it's there, even the family of Iman?" No, I said, no one knew.

"Well, then," Baba said as he settled in for the show, "no problem."

Our problem lived three doors down the block. My neighbors' son, Mezher, was a young Baathist who raised pigeons on their roof and played a popular game with them. Mezher would twirl a block of wood tied to a line of cloth, stirring the pigeons to fly up and circle around it. When the game went well, pigeons from other roofs joined in, making larger and larger concentric circles. Pigeons being stupid birds, some of them inevitably got lost and wound up on the wrong roofs.

Seven months after I bought the dish, Mezher came over to retrieve a pigeon that had landed on our house. I was at work. Iman, a trusting soul, let the neighbor up to our roof. When Mezher lifted the tarp, he knew he had struck gold.

The next day I came home for lunch and a midday nap. I changed into my *jalabia,* put on a tape of the great Egyptian diva Oum Khaltoum, and dropped off to a pleasant sleep.

It might have been the slamming door that woke me, or the footsteps in my hallway—or the presence of four men in my bedroom wielding drawn revolvers.

"Stay in your bed—don't get up!" one of them shouted. He was a large man with a cruel mouth and a puff of blond hair combed high—another Tikriti. As one of his gang ran his hand under my pillow to check for a weapon, he trained his gun between my eyes and said, "Are you Mohammed?"

"Yes, I am Mohammed. Who are you?" I corralled my thoughts and weighed two possibilities for these men in civilian clothes: robbers or security. Robbers would be better—you could work out a deal and it was only money, after all. But if they were security, you wouldn't know who sent them or how far up it went. You might find yourself dumped in a well and that would be the end of it.

The blond man ended my suspense: "I am Captain Fares, Special Security. And you are a spy! You will be hanged!"

My heart dropped. I was dealing with the elite, the equivalent of the FBI—the very top of a state security apparatus that employed one of five Iraqi males. They would know everything about me; they would know what kind of tea I drank that morning.

But there was no point in showing my fear, and I said, "What do you want? What is your evidence? I am a lawyer!" My credentials did not faze them. As the others pulled me from the bed, Fares stepped back through the doorway, where Iman was holding baby Abir. He wanted me brought out to him. I resisted, demanding to see a warrant and their ID cards. These were my legal rights, but the captain and I both knew that rights made no difference when you were up against the regime.

Fares looked like he wanted to hit me, but he lowered his hand and said, "We'll teach you your lesson at headquarters."

Another man said, "Can he change, boss?"

"No, we'll bring him in as he is."

Outside I saw the dreaded Land Cruisers: one for me and one for my satellite dish, which three men with assault rifles were dismantling from my roof. *Quite a display for the neighbors,* I thought, with me in my *jalabia* and slippers. They put me in the backseat with an officer on either side, and then came the blindfold. Up front, next to the driver, Fares was still seething over my insubordination: "When we get there, we'll show you what security is. You don't know what kind of trouble you're in."

We drove a long time; we were not going to the local precinct house. I had plenty of time to consider the pain ahead of me, and my infant daughter, and what this scandal would do to my law practice. My thoughts ran riot.

After maybe forty-five minutes, the car stopped. I was led into a building and down a hall, where a new voice called out, "Did you find the dish at his home?"

Fares said, "Yes, Major, we found it and brought it with us."

Then the major said, "*Welcome* . . . Bring him in and sit him down."

Fares said, "Let us give him a flogging before he talks to you."

"But he is so weak and small, he won't be able to stand it."

"He's not so weak, sir—this guy has a black belt." As I'd suspected, they indeed knew everything.

"Is that a fact?" the major said. Then he asked me: "Have you had dinner?"

I said, "No, I haven't had any."

"Then give him dinner."

They pushed me down a flight of stairs, into a dank,

foul-smelling place. They tied my legs at the ankles with a rope. Then they went to work, whipping me with a lash that I thought would break my ribs. I lost my balance and fell to the hard floor, banging my head, and they kept on whipping. This was not manly, I yelled, to beat someone tied up. In response they worked harder, zeroing in on my neck. I heard someone screaming, and I passed out.

I woke up to a flood of water in my face. They had taken off my blindfold, but the person with the pail was gone before my eyes could focus. My legs and hands were still tied. I felt dizzy, nauseous. I hurt all over, and each breath made it worse. I tasted the salt of my blood.

I was lying in the middle of large, shadowy room, lit dimly by one naked bulb. The walls were painted red and black: torture colors. There was dried blood everywhere. To one side I saw a soiled mattress in a metal frame, plugged into a pair of electric terminals. Next to it, a bed of nails. Overhead, screwed into the ceiling, were metal hooks—two with chains, two with ropes.

Turning my head, I saw a chair with Fares's cotton jacket, laid aside to stay clean while he labored. Then the most chilling sight of all: a table with a blank pad of paper, ready for the confession when I could take no more. Ready for the signature that would send me to my death.

Two hours—three? four?—later, my guards came for me. When my legs buckled, they dragged me up the stairs and into an office. A bald, stocky man stood up from behind his desk and said, "Now we can proceed." It was the major who'd offered me dinner the night before. His name was Akram. On a VCR by his desk they had piled my movie collection: *Return of the Dragon,* the four *Rocky* movies, *Godfather* and *Godfather II, The Yellow Rose of Texas,* a home video of Abir's first birthday party.

Akram said, "Where did you get this satellite dish, and

who gave it to you? Who are you working for—Israel? Kuwait? The United States?"

I said, "I am a lawyer working for my country."

"Don't be stupid," Akram said. "We have seen your house, your car, your possessions—I am an officer twice your age, and I do not have the things you have. You did not make all that money as a divorce lawyer."

When I told him the money came from my family, he shook his head, unconvinced. Then he said, "I know that you've had some dinner, but perhaps some tea would help your memory—would you like some tea?" He nodded to a burly young agent. "Bring him tea."

The roundhouse slap knocked me down. Akram smiled at me; he enjoyed his little jokes. He said, "Now, do you want some more dinner?"

"No," I said quickly, from the floor. "No, I don't want dinner."

"Then you need to answer my question. Where did you get the dish?"

I had made myself a promise—that whatever I did, I would not betray my friend Serwan. But I had to say something: "I bought it in Baghdad."

Akram said, "These things are not sold in the souk. Who sold it to you?"

"I don't know his name."

"That doesn't make much sense, does it? It seems that you need another dinner, after all."

As Akram waved his hand in dismissal, and his aides closed in on me, my terror cleared my head. "Wait!" I said. "My uncle is Ali Khayoun al-Nasiri, the counselor to Qusay." Why hadn't I thought of that before?

The major said, "Are you sure of that? I am going to call Ali Khayoun. If you are lying, I will add up the seven digits of his number, and that is how many beatings you will get."

He waved his hand again, and the burly one grabbed my hair to lift me to my feet. Only this time Akram said, "Don't do that, do it properly." The game had changed—or had it?

In the torture room I found two bearded men, a father and son in traditional garb. They had been charged with selling arms to the *mujahideen*, anti-regime guerrillas. As I watched, Captain Fares made them strip to their boxer shorts. With their hands tied from behind, they were hung by their arms from the metal hooks, their feet dangling above the floor. The torture began with wire cables, the same tools used on me.

Soon the men were screaming: "Stop, for God's sake!" They beseeched Ali and Hussen. They called on the Prophet Mohammed.

Fares said, "Oh, you want Mohammed? Let me call him—Mohammed!"

An officer who shared my name said, "Yes, sir?"

Fares handed him a cable and said, "Come and help us, Mohammed."

The men shrieked, "No more, we beg of you!"

Fares said, "Tell us about the men you sold those guns to."

The father said, "We didn't sell them underground—we were selling them to the tribes, just the usual business. Please believe me!"

By then the son was unconscious. They cut the father down and dunked his head in a barrel of water. Then they laid him down on the confession table. Using a bottle opener, Fares removed his fingernails, one by one.

The old man's screams rang in my ears. It wasn't hard to see what Akram was doing. Because of my uncle, he could not take a free hand with me. He needed to extract my confession by subtler means—by breaking me psychologically.

His strategy was working. I was ready to tell him anything, even though I had nothing to tell. Far better to die in peace, with a clean shot to the head, than to go this way.

I heard a noise on the stairs. It was Akram, who told Fares, "You'll have to hurry this up. We have one hour to get this confession if it's going to reach Baghdad in time."

Fares signaled to his aides, who smeared the groaning old man with animal fat. They brought in a German shepherd and let go the leash, and I flinched as the dog flew past. The shepherd had been trained not to rip through the jugular or any vital organs. Instead, it gnawed at hands, arms, thighs. It was an atrocity, a nightmare, but Akram got what he wanted. When they pulled the dog off, the old man signed the paper on the table.

I lay in that room with those two mangled bodies for hours past counting. It was the next morning before Fares came for me one more time. Had my uncle come through? Had Akram reconsidered? Or would the satellite dish be my doom?

As we neared the major's office, I could hear a familiar voice—it was Baba! He was sitting with my brother Ahmed, and Akram, and a man I didn't know. An aide served tea and coffee. Everyone seemed like old friends.

Akram said, "Come in, Mohammed. I want you to meet my brother."

The story was told. Akram had reached Ali Khayoun, who confirmed that he was my uncle—but added that if I proved to be a spy, he would shoot me himself. (I knew that he was not exaggerating.) Meanwhile, Baba used his connections to find out who was holding me. A business friend of his was close to Akram's brother, a pilot in the Iraqi air force. My father and brother drove the seven hours from Nasiriya to Tikrit, met up with the pilot, and drove the seven hours back with him to vouch for me. For his trouble, Baba paid the pilot $4,000.

They all laughed as Akram tore up the confession that I had not quite signed. "You know," he told Baba, "your son is quite the troublemaker."

"Yes"—Baba sighed—"he has always been a painful child." They laughed some more. I felt silly sitting there in my *jalabia,* but I cannot lie. It was a great moment in my life.

Akram said, "Excuse me, Mohammed, I have forgotten my manners. Would you like some tea?"

I very much wanted some tea, but glanced at the aide and thought better of it: "I'll have a Coca-Cola, please."

After I'd changed into the sweatsuit they'd brought for me, Akram said, "Listen, Mohammed, when you leave here and people ask you what happened, tell them that someone left the dish at your house. Tell them you were just keeping it as a favor, that you never used it."

And I thought: *Even monsters must keep up appearances.*

To honor the pilot's visit, Baba invited Akram and Fares and their whole headquarters to a dinner at Nasiriya's best hotel, with four lambs slaughtered for the occasion. While Special Security feasted, I soaked my wounds in my tub. Baba was right. I was the painful son, and it seemed to be catching up with me.

CHAPTER THIRTEEN

A BIR WAS twenty-one months old when it started, a deep cough that shook her whole body. Our family doctor called it an inflammation in her right lung. It should pass, he said, but it did not. The next doctor said it was whooping cough and ordered antibiotics. Three days later, Abir was worse, with a high fever. Over the next four months, I took her to Basra and then to Baghdad—to seven different doctors, all well respected. Each one had a different diagnosis. None could make my daughter well.

Whenever I took Abir to yet another clinic, I would see how the sanctions had punished the people of Iraq. With rice cut back to two kilos per head per month, malnutrition was epidemic. I saw children so weak and pale, with sticks for limbs. They were given a vitamin solution, a poor substitute for food.

Iraqi medicine had once been the envy of the Arab world. Now we lacked basic drugs and equipment broke down for the lack of parts. Diarrhea spread among the young; there was cholera and typhoid in the countryside. Once we sent a driver to Jordan to get medicine for Abir and some others. He took our money and stayed there.

I applied to the Ministry of Health for a travel permit. My hope was to take Abir to Amman or Damascus, where facilities were better, and from there, if need be, to Europe. Though the trip would have been at my expense, the government turned me down. If they let people like me out of the country, why would we come back?

Eight months after the sickness began, we went back to Baghdad, to the place that treated the Baath Party elite: Adnan Hospital. For the next four months, Abir would go in for twenty days and come out for ten, as Iman and I split the bedside shifts. They tried every experimental treatment. A visit from French and British doctors fell through. Abir kept getting sicker and weaker. She lost weight. It hurt her to inhale.

Finally the hospital's chief surgeon told us, "The tuberculosis has infected the entire right lung. It won't work any more. It must be removed."

Though I wasn't trained in medicine, this seemed illogical. If Abir had tuberculosis, why was there no blood in her cough? But Iman and I were desperate. We were watching our small daughter die in front of us and had no place to turn. If the top people in Iraq said they knew how to save her, how could we say no?

The surgery was radical. They cut Abir from the middle of her stomach, around her back, and up to her shoulder. When the surgeon came out, he told us, "Everything went well. She should live a normal life."

I asked him, "What did you see in there? Did you find the tuberculosis?"

He walked away without another word.

They were hiding something, I thought. I asked a doctor friend, a cancer specialist, to inspect the lung in the hospital lab. He came back tight-lipped and somber—out of sadness,

I thought. He said, "The lung was badly damaged. They did the right thing to remove it."

There were complications, and for four days Abir was in critical condition. We almost lost her. When she came out of it, I tried to put the ordeal behind us. I still had doubts about her treatment, but what was done was done. It would do no good to sue the most prominent doctors in Iraq.

Abir was not the same child after that. Her appetite flagged. She tired easily and seemed to catch every cold. When I bought her a tricycle, she tried it once and then left it alone. It was too strenuous for her.

Sometime after the operation, I went out for a few drinks with my doctor friend. When our talk inevitably turned to Abir, he said, "You have always been kind to me, and there is something I need to tell you. I never saw Abir's lung."

I was stunned—how could this be?

When he got to the lab, my friend said, the surgeon told him the facts. They had lied to me about the tuberculosis. Abir's problem was a pistachio shell she'd inhaled all those months before. The shell was easily removed. The lung itself was fine.

"Then what *happened* to it?" I could hear my voice, but it seemed to come from far away.

There was another girl in Adnan Hospital, the daughter of a high Baath official, my friend said. She had lung cancer and needed a transplant. And there was Abir—an anonymous Shiite child of no consequence. A perfect donor.

"And they took her lung." My friend was tormented. "They took Abir's lung, and they threatened if I told you . . ." Where our earlier doctors had been incompetent, these top surgeons knew exactly what they were doing.

What could a father say to this? I felt numb. The story was too grotesque. In any country but my own, it might have been impossible to believe.

I went to Adnan Hospital for an explanation. They referred me to the Ministry of Health, where a clerk said that my daughter's file was closed. There was no appeals process in Iraq, no freedom of information. I was at a dead end.

One day I would take my revenge against the Baathists. It was a loose threat, for now, but I swore it would not be an empty one.

Until then, I could only care for my daughter, and love her. And try not to weep when I put her to bed and saw that long, pink scar.

In 1999, four years into my practice, a woman came to me from Basra with solid grounds for divorce. Two months into the case, all was going well when her husband called and told me to drop it. I said that was impossible. A few days later, there came a pounding at my door. As I opened it, a man grabbed me by the shirt and said, "Look, you. If you don't drop the case, we will drag you in the streets. You will disappear."

It was the husband, a captain in defense intelligence. I took his threat seriously. Abir was sick at the time, and who would take her for treatment if something happened to me?

The next day I called the woman and quit on her. Everyone in the courthouse knew about the case, and I felt humiliated. I had broken my professional oath. Then again, the oath was a sham. In Iraq, the rule of law mattered little next to the rule of Saddam.

I never took another case. I closed my practice then and there, stuck my diploma in a drawer. It spelled the death of my mother's dying wish, but I was not meant to be a judge.

I found a storefront gym and opened a school for kung fu. Within a month I had ninety students, doubling my income as a lawyer. It was an honest living, and a safer one, too. I could go about my business unnoticed. For a long tongue like me, perhaps that was just as well.

By the time I switched careers, I was the top-ranked kung fu competitor in Iraq. I had won the semi-annual national championship in 1996 and 1998, as well as major tournaments in Jordan, Lebanon, and Egypt. I was headed for the Asia Games in China until Uday Hussein scratched all Shiites from the list, for fear that we might defect.

We called kung fu "the Game of Death." We used no protective padding and only the lightest leather wraps for our fists. Matches between black belts were especially treacherous. I had broken my arm in the ring, and I considered myself a lucky one. Twice I had seen contestants killed.

In 2000 I came again to Saddam Arena in Baghdad to defend my title. By the second day of the tournament, only three other players were left: Waleed, Samir, and Adnan. The four of us were good friends. When we met for breakfast, Waleed said with a smile, "God willing, I will beat you all, and then I will have two things to celebrate." He was to be married later that week.

Waleed drew the first match that afternoon against Samir, and seemed distracted. Early on, he left his right side open and Samir found his rib cage, unblocked. By the sound of it, I knew the kick was lethal when it landed. Waleed collapsed, blood pouring from his mouth. They raced him off in an ambulance.

In the second match, Adnan beat Samir; he would face me the next day in the final. As I met with friends and relatives outside the arena, we heard gunshots. "It's Samir!" people said, running from the scene. "They killed him!"

Waleed had died en route to the hospital. Though it was a great sorrow, no rules had been broken—this was the chance we all took. But Waleed's brothers were crazed. They picked up their guns and lay in wait for Samir, and now there were two deaths. (Soon there would be more. Enraged by the loss of a top player, Uday had Waleed's brothers executed.)

I went to our training gym to clear my head. Iman and Ahmed came to see me, saying they had a message from Baba: I must withdraw from the tournament, or else be disinherited. I looked at Iman's eyes and knew that my brother was lying—my father had said no such thing. But they were frightened by what had happened to Waleed and Samir. They wanted to protect me.

There was no shame in feeling fear before a match. Waleed had made a bad mistake, yet it could happen to anyone. It could happen to me. If I backed out now, I would be finished as a black belt. I would lose my dignity and my livelihood. What would be left of me?

I learned this from Ciao Lee: *Never fear to act—and when you act, never be afraid.* Yes, we all felt fear, but the task was to master it, to set it aside before the match. If I brought it with me, I would surely lose the title. And maybe a lot more.

The next afternoon, as I walked to the ring, I saw the table full of trophies to my right, with a jeweled belt for the champion. I saw the medical emergency team to my left, its ambulance at the ready. I knew the odds were good that I'd end up on one side or another.

As I stepped onto the mat, though, I saw only my opponent. I was quick to the punch that night; I had never felt sharper. I won the belt.

Martial arts were widely followed in Iraq, and my victory was a big story in Nasiriya. Our local newspaper, *al-Baath Riadi,* called me *battal,* or "champion"—which is also our word for "hero."

In my own eyes, though, I was not a hero for *winning*

that night. I was glad to have beaten Adnan and kept my title, but those were not the hardest things. The hardest part, the heroic part, was to walk past the ambulance and into the ring, knowing my friend had been killed the day before. The hardest was the *risking,* win or lose, no matter what. To take the chance, and not look back.

PART THREE

CHAPTER FOURTEEN

ON MARCH 27, the day I saw the young POW and set off for the Marines, I had a rough idea of where to go. Judging from the direction of the shells and what we'd heard from the refugees, U.S. forces were camped to the east of us. I guessed their distance as four miles as the crow flies, more than that through the alleys and fields between us. The sun would set at six-twenty. That left me four hours to get to the Americans before dark—before the guerrillas came out, and people started shooting at anything that moved.

It was cool and clear that day in Nasiriya. I set off walking between houses, avoiding main streets, moving parallel to the south bank of the Euphrates until I reached some abandoned farmland. It was dotted with palms and clumps of grass, cut by small streams and canals. There was an occasional stray cow, without much of a future. Iraqi soldiers were expert rustlers, roasting their catch over open pits. I felt vulnerable here; I would have to be alert.

I reached another stretch of houses where the fighting had been close enough for the residents to evacuate. Every now and then I spotted someone from my neighborhood, or a person I knew from downtown. Some were camped by deserted houses or amid stands of trees; others loitered on the streets. They seemed to be doing nothing, but they were there for a purpose: to watch, and inform.

"Hello, *austad!*"

I nodded and smiled at a group in green Baath uniforms as they unloaded ammunition from a pickup. The one calling to me was a stocky, middle-aged man who had taken my kung fu class. He'd been sent to me by the emergency police unit, where he worked part-time. I could feel his stare as I kept moving east, and I knew what he was thinking— there was nothing in that direction but Americans. Soon the word would go out: that Mohammed al-Rehaief was consorting with the enemy. By that evening the rumor would have a life of its own.

From a distance I saw a gigantic work party, easily two hundred people. The fedayeen were dragooning any passersby to help them dig their foxholes. I swung south away from them, toward yet another residential area—and smack into an Iraqi army platoon in green-and-yellow camouflage.

The soldiers asked where I was going, a fair question. I told them, "My house is over in Zenawiya. I need to get my money and medicine for my daughter."

One soldier said, "Are you out of your mind? Go back, go back. Do you want to die?" We heard more explosions, not far off. The soldier said, "Either you leave or I am going to shoot you. We don't have time for you here."

As I retreated, the shells closed in. The air filled with black smoke. I jumped into a hole and huddled as low as I could. A near miss spattered my head and shoulders with clods of dirt. I lost most of my hearing. I thought I was done for.

I faintly heard a soldier calling, "Major Ayoub says to stop shooting, it's giving away our location."

The bombardment lulled. I looped back north and then east again, to a sandy area with middle-class homes and scattered palm trees. I passed a charred Iraqi tank, still aflame— I had reached the front. At an irrigation ditch I ran into more soldiers, a full company this time. They were dug into trenches or prone on the ground.

"Stop!" one of them yelled up at me. "Are you Iraqi?"

I said, "Yes, I am Iraqi. What's your problem?"

"What's my *problem*?" He was a big, leathery man, the kind who liked to push people around. "What are you doing here?"

I used a new story, something that might impress him: "My brother is a fedayee and they told us he was injured near here. I'm going to see him."

The soldier said, "Forget about your brother, go back and check your mother. The Americans are right behind us."

I knelt next to him in the foxhole. "I want to talk to your major."

"Do you know Major Talib?"

Thank you, I thought. "Yes, I do."

"He's very busy now."

"I'm going to wait."

The soldier said, "What do you want from him?" A shell smashed into a date palm just behind us. "We wish that *we* could run away, but they'd execute us. I think you're nuts."

Ten yards away from us, a beefy officer left his trench for a latrine. He had stars on his shoulder and an eagle on his cap. I took my gamble, knowing I could not afford to be wrong: "Major Talib! Major Talib!" The officer barely glanced my way.

The leathery soldier said, "He didn't recognize you. Why are you lying?" He began to sound threatening.

I scooted out of the foxhole and ran toward the officer, shouting his name. The soldier lagged behind me, shouting, "Sir, do you know this man? He said he knew you!"

The officer said, "You're an idiot—I've never seen him before."

"Wait a minute, Major Talib, let me explain!" If ever I needed to be a clever lawyer, this was the time. "My uncle is Major Ayoub—he says hello."

"Ayoub is your uncle?" Talib told his men to back off. "What do you want? Quickly, I'm in a hurry." He squatted; I copied him.

It occurred to me that an officer might know where the fedayeen were—and where they weren't. I fell back on my first invention: "Our house is in Zenawiya. My money and my daughter's medicine are there. I went to my uncle, but I couldn't find him. He told me before, 'Go to Major Talib if you ever need some help.'"

Talib was a thick-featured man in his forties with a guttural voice and a boorish way about him. He said, "Didn't your uncle tell you that this area was cordoned off? You'll be killed by our people or the Americans, one or the other. You're still a young man—go back."

"But I need the medicine and the money."

"You can get new medicine from the hospital."

"But what about the money?"

The major looked at me with new interest: "So you like money, do you?"

I said, "Doesn't everybody like money?"

Leering at my wedding band, Talib said, "True enough, but I like gold even better."

I hated to lose the ring, which was inscribed with Iman's name. But taxes were high in a war zone. When he saw me tugging at my finger, Talib said, "Not in front of the men." Half a dozen soldiers, flat on their stomachs, had their rifles still pointed at my chest. The major and I duckwalked to the latrine. As I pulled off the ring, Talib spotted my waterproof Japanese watch and said, "Give me that, too, to remember you by."

A sniper's bullet hissed by our ears, and we hunched lower by the stinking trench. This was a tense transaction, because the watch was my window on time—and my life insurance. It was close to four o'clock. If I wasn't close to the Marine camp by six, I would need to turn back. Much later than that and the Americans would see me before I saw them.

I had little leverage, however. If Talib insisted, I would hand over the watch. I said, "I would love to give it to you, but it isn't mine."

The major said, "Who said you're going to live through this? You might as well give it to me." After straining to fit my wedding band on his ring finger, he screwed it on his pinkie with a grunt.

"We'll see," I said defiantly. We locked eyes until he looked away. Maybe Talib wanted to avoid trouble with my uncle Ayoub. He let me go.

I moved on through some low-lying hills, pausing when the shelling got bad. In twenty minutes I had reached Zenawiya, a district of small stone houses. The streets were a mess of crushed asphalt. Half the homes were demolished, with not a soul in sight.

Cutting through somebody's garden, I met up with a band of green-uniformed men with a patch on their left breast: the Dome of the Rock and a motto, *Long Live the Homeland*. These were the Jerusalem Brigade, an elite mercenary unit of retired military officers. They had been mustered by Saddam for his goal of "liberating" Israel. Back in January, the last time I had seen Uncle Ali Khayoun, he told me that Saddam had won safe passage for the brigade through Jordan and Syria, and soon they would be ready to strike.

Of course, Saddam was a famous blowhard.

One of these men told me, "Get inside your house."

I said okay and walked around the block—only to meet the same group again. To avoid suspicion, I went to the nearest ruined house and sat on a rock in its garden. I wrung my hands and muttered, "God damn the Americans!"

One man came to me and said, "May God compensate you. Did somebody die here?" I buried my head in my hands and pretended to weep. The man said, "Don't worry, you have been avenged. Yesterday four people from here went to the Americans on a suicide mission."

I continued to grieve, not looking up. When the street cleared, I went east till I reached the last open area before the Marines—a marshy no-man's-land, a few hundred yards south of Ambush Alley. I heard sniping from both directions; Iraqi mortar rounds sailed over my head. I hit the ground and watched a helicopter buzz the mercenaries in Zenawiya. They would be too busy running to shoot for a while, or so I hoped. I sprang from one bush or tree to the next, keeping covered as best I could. I was parched with thirst, weak in the knees. Mosquitoes dogged me. The sun was low. *Time is like the sword*, I thought. *Too late, and it will cut you.*

I could hear the Americans to the east—the rumbling of tanks, the artillery and automatic weapons. I slowed to a hobbling trot. Each step was a small torture—would it be my last? A quarter mile from the activity, I gave in to exhaustion and collapsed in a thatch of wild grass, gulping for air.

The horizon turned a deep pink. *Red sky at night, shepherd's delight.* It would not rain tomorrow, I thought. Then again, I might not be around to enjoy the weather.

A spray of bullets threw up sand all about me. At dusk

the Americans combed the front to discourage intruders. The sweep was fierce. When they finished (or reloaded), I got shakily to my feet, dashed forward with what energy I had left. I pushed to the top of a low hill and ran down it with my hands in the air. Now I could see the American tanks, two hundred yards away. I could only hope that the little hill might screen me from any Iraqi with binoculars and an AK-47.

Looking back, I can see that this was a bad idea. I was behaving like a textbook suicide bomber. The Marines must have jumped into their foxholes and taken aim. I heard the faint click of their rifles—

"Don't shoot!" I cried, grasping for the words. "I am a friend!"

I heard a voice call out, "*Stop!*" I stopped. There were more instructions that I could not understand. I wished I had been a better English student. Recently we'd heard of an Iraqi civilian, snared in a firefight, who had muddled some American's order and been shot.

Then I saw the bodies, sprawled to the right of me in the sand. There were two men on their faces, two women—one heavy, one thin—on their backs. The suicide crew from Zenawiya. Rattled, I stepped away from them.

"You stop right there, you son of a bitch!"

I froze. I had never heard that last term before—was it a word for surrender? I forced myself to smile and shouted, "I am a friend! I am a son of a bitch!"

"Just don't move," the voice said. After some trial and error, I spread my legs as far as I could. I made out some movements in the twilight, until the flashlights blinded me. I lowered my eyes by reflex and the voice said, "Raise your head." And then: "Keep one hand up. With your other hand, pull up your shirt from your chest." Confused, I lowered my hand to my waist. "*No-no-no-no-no!*" went the chorus. *"From your chest!"* I corrected myself in time and showed that there was no bomb.

I heard them moving closer. Someone said, "Lie on the ground!"

I said, "I don't understand."

"Shut your mouth!" He sounded young and nervous. The men inched back, until one of them had an idea: "Lie on the *floor!*"

That time I got it. I dropped to my knees and then the sand, facedown. In an instant I was surrounded. One Marine stuck his boot on my back, others on my ankles and hands. They searched me quickly and thoroughly from the neck down, taking my watch and wallet, binding my hands and feet with plastic coils. They blindfolded me and threw a hood over my head, and carried me off on the run.

A hundred yards or so later they put me down, my back against a wall. I sat there for a long time, in total darkness, trying not to jump at each explosion. Now and then someone shone a light on my hood. I heard one man talking on a phone. Another said, "Be careful; don't say anything near him."

I did not anticipate such rough handling. Was there worse ahead? I had come here of my own free will, yet they were treating me like a prisoner. I had heard that Americans tortured Iraqi prisoners for information.

At last they untied my feet, and for fifteen minutes we walked over choppy terrain. I stumbled several times. Next came a bumpy truck ride, with the crash of heavy guns all around us. The plastic tie dug into my wrists. One of the soldiers asked me, "Do you like America?"

I said, "Of course," and he found that funny. But the one sitting up front told him not to talk to me. The truck stopped. I was hoisted out, led into a room, and ordered to sit on the floor. I heard someone go off to get the sergeant, who soon arrived with a translator.

"*Gowa*," the translator greeted me.

I said, "Oh, you speak Arabic? Where are you from?"

The translator said, "You don't ask the questions." When the sergeant inquired about my dialect, he said, "Iraqi."

The sergeant asked what I was doing there—was I a soldier or civilian?

And I said, "Before I answer any questions, you have to take off this hood and blindfold."

The translator said, "Oh, we *have* to, do we?" After hearing the translation, the sergeant said, "Do it." My eyes had been closed for so long that it took me a while to adjust.

The translator said, "Okay, now you can see. Tell us."

I said, "No, first untie my hands. I am not a prisoner of war. I am coming here to help you, and this is not good treatment." I could see them standing in a half circle around me: a staff sergeant, whose name was John; the translator, Abdul Rahman, a stocky young guy with a shaved head; and three guards with rifles. They wore khaki camouflage rain jackets and baggy pants, protective suits against chemical warfare. From al-Jazeera, I knew they were Marines. (I later found out that the translators were Kuwaitis. They'd been given Marine uniforms to protect them as POWs if they were captured—though if the fedayeen did the capturing, it would not matter.)

As I sat on the tile floor of some classroom, the sergeant ordered two more guards inside and removed the plastic coil. Rubbing my wrists, I said, "Now bring me water to drink, and a chair. I don't want to sit on the ground."

When he got the translation, the sergeant smiled. They brought me a chair and a bottle of water. "Now *talk*," he said.

In English, I told them, "I am a friend. I have important information on the fedayeen. I am a son of a bitch."

"What?" The sergeant was taken aback.

Abdul Rahman said, "Who taught you that?" When I explained, he told me to switch back to Arabic.

After some back-and-forth, the sergeant asked Abdul Rahman, "What do you think? Does this guy know something or not?"

The translator said, "I'm not sure about that, but I think he's dangerous."

As they left me, I felt discouraged. I was prepared to be tortured, but not ignored. What was going on?

Two new counterintelligence men—one in khaki camouflage, one in tan—entered the classroom. This pair was friendlier, shaking hands and introducing themselves. Staff Sergeant Corey had a quick smile and short blond hair, combed back. Khalifa, his translator, was older and heavyset, with black hair in the same style.

Corey said, "We're here to talk to you, and what you tell us will determine what happens to you. We need you to be very honest with us."

I relaxed a notch—and remembered why I was here. I said, "I know about the American POW, she's in the hospital."

"Which hospital?"

"Saddam Hospital."

I was puzzled when neither man seemed to take me seriously—they actually laughed! The sergeant said, "I happen to know there is no such POW. Look, just forget about that and tell us the truth. Who sent you?"

We were getting nowhere. I'd been unsure about mentioning the suicide bomber in the ambulance—what if the fedayeen had already done the job, and the Americans thought I was a party to it? But I needed to prove myself. I said, "I'm going to tell you something and then I will stop talking. Time is flying by." I told them about my neighbor Abbas and added, "Better for you to stop this before it happens, and then you can ask me questions and save the POW."

Speaking my language, Khalifa said, "What you are saying is logical, but these Americans are stupid people. I am your Arab brother, and I want to help you. Talk to me, just between us. This one"—he nodded at Corey—"doesn't understand. If somebody is sending you here to do something, just tell me. We can get you out of your problem."

With no food for twelve hours, I felt light-headed. If these were games, I could not play them. I said, "If you are logical, think about this. I risked my life just to be here. Of course I am telling you the truth."

Corey said, "Very smart answer," which surprised me, because I didn't know he understood any Arabic. "So you say there's a POW?"

I nodded eagerly, and began to tell them about my morning at the cardiac unit. Before I got far, however, a knock came at the door. Staff Sergeant John entered first, ahead of two guards who flanked a hooded Iraqi, with Abdul Rahman bringing up the rear. John approached me with a nasty smile and said, "What are you *doing*, Mohammed?" (*How does he know my name?* I wondered. *I haven't told anyone.*)

And Abdul Rahman said, "The truth has come out. This fellow got a lesson from us and confessed to everything." He asked the Iraqi: "Mohammed is with you?"

The hooded man said, "Yes, he is with us," and they removed him from the room.

Corey looked at me sadly and said, "See, we trusted you, and now you turned out to be untruthful. We'll give you five minutes to think about this. If you don't tell us the truth, you will regret it."

They left me alone with the guards. The five minutes became an hour before everyone came back. They pulled a table in front of me, with pencils and pads and a tape recorder. Corey closed the door and said, "Start talking."

I said, "I have nothing to talk about. The truth is the truth."

Corey said, "How come this guy knows your name?"

I said, "I don't know."

They brought me a glass of water, filled to the brim, and directed me to hold it at arm's length. This would be my lie detector—a tough test, given my fatigue. When they asked a question, I must look at them as I answered.

"All right," Corey said, "let's try this again. How come this guy knows your name?"

The water did not spill, but I had no answer for them. Could I go to the bathroom? Abdul Rahman said, "Oh, you want to get some time to think of an answer, right?" I admitted he was right. "You see, we can read your mind."

I said, "Time is passing us by. This is the last time I will tell you about the ambulance, or we will miss our chance. The information will be worthless. If you believe me, you will be the ones to gain. If you don't, you will lose."

Corey said, "All right, this is the last question, and then we'll get the people with the guns out of the room and we'll talk. What made you come here and risk your life?"

I replied, "What are *you* doing here?"

Corey said, "This is my job. What about you, Mohammed?"

I said, "This is not *my* job. This is my duty."

"Your duty?"

"My humanitarian duty."

Corey said, "That's great, but you're a Muslim, a Shiite. Why would you help an American?"

I looked at him squarely. "Because each life is precious. Even an American's."

They sent out the guards, fed me some cakes, and got down to business. They took down the details about Abbas's ambulance—an '89 model, French—and relayed them to their men in Mansuriya. Over the next five hours, on and off, they questioned me about the POW. Though I did not know her name and had barely seen her face, I told them about the fedayeen colonel and what I could recall of the hospital layout. They sent this information by cable to their headquarters in Qatar.

We went at it until two-thirty the next morning, when I could no longer will my eyes open. The Marines treated my blisters and set out a sleeping bag on the classroom floor. I was grateful to lie down, but sleep came hard. I fretted about Iman and Abir. I'd been seen moving toward the Marines, which marked me as a dead man in Nasiriya—and my family right beside me. I had no way to reach them; our cell phones had been useless since the war began, and the landlines went out soon after. There was nothing to do except wait to go back.

It was still dark when someone woke me and said, "You were right." Abbas's ambulance had come to them at four that morning. The Marines were ready for him, and no Americans were lost.

There was no sleeping after that. I was taken by jeep to a nicer place, an office building. They brought me orange juice and cakes and my first "Meal Ready to Eat": cold chicken soup. I left the MRE but ate the rest. Now that I had the Marines' trust, I asked about the hooded Iraqi and how they knew my name.

Corey said, "That was easy. The Iraqi works for us, and we got your name from the ID card in your wallet." He returned the wallet to me.

Around dawn they brought in a tall young man, a private who seemed highly embarrassed. He said to me, "I'm so sorry, sir. I want to apologize."

When I asked why, Corey explained that this was the sentry who had called me the bad name the night before. "You are a good man, a brave man," Corey said. "You are no son of a bitch."

I began to understand. After twelve hours with the Americans, my English was already improving.

CHAPTER FIFTEEN

THE SHELLING RAGED on from both sides into the morning of Friday, March 28. At around seven-thirty, Corey thanked me for what I had done. The roads into the city were precarious. I could stay at the post for now as their guest.

Then he said, "But I have to ask you—can you help us rescue our Marine? We need more details about the POW and the hospital. It's a dangerous mission. We can't make you do it. But if you want to volunteer, we'll explain how you could help."

My job was not finished—it was that simple. I said, "That's why I came here, to help you get the woman back."

Corey and his translator traded smiles, and we went back to work. They needed to know *everything* about Saddam Hospital. The troops and vehicles around the perimeter—and were there police dogs? The entry points into the compound, the sites of the outbuildings. The hospital exits and stairways and elevators and hiding places. The layout of the cardiac unit. The number of soldiers and fedayeen inside the building, the weapons they carried. The dimensions of the roof—was there room to land a helicopter?

Then came the hard part. I had to penetrate the POW's room and make sure she was responsive, that her speech and hearing were okay. Was she on IVs? Could she be safely moved? I needed to bring back more details: the color of her eyes, her hair. They wanted to be absolutely sure who she was.

And finally, if possible: What was her name?

(Later on, I realized that they'd had a good idea of the POW's identity. At the time, only two female American soldiers were unaccounted for. When I told Corey of the brown-skinned soldier I'd seen paraded on the roundabout, he must have deduced it was Lori Piestewa. But he did not tell me this, nor did he name the other missing woman. He wanted me to go in with fresh eyes.)

Corey said, "We want you to protect yourself. You can't write anything down. Make your forehead a sheet of paper and your finger a pen."

I agreed to everything, with one condition: that Iman and Abir be protected. I told Corey, "You don't know what it's like in Iraq. If they don't arrest me, they will get my family."

Corey brought his fingertips together and said, "Your security is like a triangle. Nothing will be revealed from our side. And on your side, we know you won't make a mistake and expose yourself. But this third side is open: your family. For you to do your work in peace, we have to protect them and close the triangle."

If my family was threatened, Corey said, I could bring them back with me, either to this place or to a second Marine camp, farther south, where the route from my house was more direct. All of them—not just Iman and

Abir, but also Baba and my brothers and *their* families—
would be safe there until Nasiriya settled down.

There was one more thing, Corey said. To help the sentries know me when I returned, I needed to wear something
more distinctive than my denim jacket and jeans. I knew
just the thing: an outfit belonging to my brother Hussen,
who stored clothes at Baba's and was closest to my size. He
had terrible taste, but now it worked to my advantage. I told
them what I would wear: a short-sleeved black print shirt
and tan checked pants.

Just before I left, a Marine officer stopped to see me privately. He asked if Corey had promised me money or
some other reward.

I said, "No, he did not."

He said, "Are you expecting to get something out of
this?"

"Not unless you want to give me a gun."

The officer told me, "No, but you're going to get something bigger than that. Live or die, you'll get a place in
history."

The Marines apologized for not driving me partway back.
They couldn't take the chance that an Iraqi might spot me
with them. I would have to go as I had come, on foot.

Corey and Khalifa walked me out of the office building
and to the nearest road. They retied my hands behind me
with the plastic strip, but loosely, so I could free myself
when I wanted. If I ran into the wrong people, I could say
the Americans had taken me prisoner but I escaped. If I ran
into other Americans, I needed to remember to pull my
shirt up and out from my chest, to show there was nothing
underneath.

Corey said, "Remember, the rescue will only be as good

as what you bring back to us. But be careful. If you can't get the exact information, don't come back. If you find yourself in danger, don't come back. And whatever you do, don't come at night."

I said, "I will get what you need. I will do everything I can."

Corey said, "We'll be praying for you."

My watch read nine o'clock when I started down the dusty road, heading due west toward town. Before the war, this road would be packed with cars. If you saw anyone you knew, you could hail them and they'd stop. You would offer them a few thousand dinars and sometimes they'd take it, sometimes not. But there were no cars now to be hailed.

I walked rapidly, uneasily. After half a mile I heard automatic weapons fire, very close—whoever it was seemed to be shooting my way. I ducked behind a mound of bricks by an unfinished store, not quickly enough to fool the American patrol car. Five Marines jumped out to surround me. It was just as before: *Raise your hands! Don't move! Pull up your shirt!*

They retied my hands and knotted the blindfold tightly. As they pushed me into their car, I kept saying, "I must talk with Corey." They told me to shut up.

We drove to a deserted house nearby, where they brought in a Saudi translator. I kept asking for Corey, but my pronunciation must have been awful. "Ah, *Corey,*" the team leader said, after I wrote the name in Arabic. He raised the sergeant on his walkie-talkie. I gave Corey the password: *Iman.* I was clear.

The leader apologized. I'd been arrested, he told me, because I'd run from them like an escaped POW. They passed me water and chocolate-covered cookies, which was all they had with them.

The Marines advised me not to go farther west—they had a little fight on their hands with the Iraqi army there. I could hook south and then west again through Zenawiya, but I'd had my fill of the Jerusalem Brigade. There was one other option: to swim the Euphrates and go straight to my father's house on the other side.

The river was wide, at least seventy-five yards across and ten feet deep. There were poisonous water snakes, and snapping turtles two feet long.

I was not the best swimmer, but these were unusual times. When I reached the southern bank, I stuck my wallet in my mouth and waded into the dark blue water. I got out on the other side, muddy and dripping, and turned left toward Baba's.

Soon I was safely in Sabba, a wealthy eastern suburb. Zigzagging between houses, I passed swimming pools and lush gardens until I saw a man I knew closing his garage door. Hajji Adnan, a furniture maker, had sold Baba every table and chair in our family's hotel. He was a widower with a young second wife and twin six-year-old daughters. His house was enormous, and he drove a white Mercedes.

Returning my wave, Hajji Adnan said, "What's your news? Who died?" It was the ritual greeting for people in my city these days. He looked at me more closely and said, "What happened to you? How did you get so wet?"

I said, "I was getting water from the river and I fell in." The businessman laughed and asked where I was going now. I wanted a ride to Baba's, but felt shy about messing up his leather interior. I told him that I needed more water—I would see him another time.

I watched Hajji Adnan wistfully as he drove down his curving, dead-end street. The sun was high; it was just after noon. When he slowed to turn into an alley that led downtown, two men in leather jackets and jeans converged on the Mercedes. They were just close enough for me to recognize them from the TV news: a team of notorious local convicts. They'd been about to be hanged for armed robbery when Saddam's amnesty for criminals released them.

I saw one of them yank open the passenger door and move inside with his weapon drawn. The other pulled the driver out, hit him in the face with his gun. Hajji Adnan grabbed onto him and refused to let go.

I heard him cry: "Help! Help, Mohammed! *Help* me!"

"Let him go!" The voice came from a rooftop, a neighbor scouting for helicopters. "Let him go! I'm calling the police!"

"*Help*, Mohammed!"

I would not reach Hajji Adnan in time, and could do little against two gunmen if I did. Maybe that is why I paused half a beat before breaking into a run, or why I went at three-quarter speed. I kept hoping the businessman would come to his senses, that he'd let them have that fancy car and go home to his wife to nurse his wounds.

But Hajji Adnan was stubborn. I was thirty yards away when the robber shot him in the head, hopped into the Mercedes, and zoomed off. I saw the slim second wife run out, the two little girls. They threw themselves on the bloody street, wailing as they cradled the body.

For the first time since my mother died, I cried.

The police had surely been called. If they found a drenched man on foot, far from his home, it might lead them to questions I would not want to answer. So why did I stay? Because I could describe the robbers; because I was ashamed. I could taste the cowardice in my mouth like spoiled food.

The police did not come. If someone had called about Mohammed Odeh al-Rehaief, *agitator,* I would have been snatched up in a breath. But homicide no longer mattered to the people who ran Iraq, unless they were the ones pulling the trigger. After waiting half an hour, I stole away.

I crossed the city line, a mile east of Baba's house. Cobra gunships were raking this part of town. The streets were littered with debris and white pickups, all evacuated. The fedayeen had fled into the local mosques, knowing the Americans would bomb around them. There were few drivers out, and the last thing I expected was to hear someone call my name over a honking horn. A white Volkswagen idled at the curb. It was Maher and Alaa, two friends who lived near my father. They looked frantic.

Maher said, "What have you done? The fedayeen have been to your house. Hassan told us to warn you not to go there." The phones might be down, but the regime's grapevine was still in working order.

I said, "But what about Iman and Abir?"

Maher said, "All I know is, they aren't there. And you better stay away, too. You know they'll be coming around to see your father."

I broke into a sprint. After the businessman's cruel killing, I *had* to get to my house—to see what had happened there, maybe pick up my family's trail. I was racing the fedayeen and they had a head start.

My home was farther than Baba's, more than two miles away, but I made short work of them. I ran west near the riverbank, finding the wall of the nearest building whenever a bomb exploded overhead. Shrapnel cascaded to the ground in waves. At Victory Bridge I took a breath and dashed out along the walkway. I made it to my side of the river without passing a single vehicle—fortunate for me, since there was nothing to draw the Americans' attention.

I was close now. I circled several blocks to the local soccer stadium, found a broken door, and walked far enough up into the stands to observe the back of my house. The street behind it was quiet, as usual, just a few kids playing by their doorway. The house that fronted on the back of mine was the one I worried about. The owner, Abu Amar, was a Baath official. His three sons were all in the fedayeen, and one of them lived upstairs. They openly envied our larger house. They would be overjoyed to catch me for a fraction of the 15 million dinars.

Since Abu Amar's front door faced the door to my kitchen, I crept around to the side. Finding the kitchen windows shattered, I used a shoe to knock out the few remaining shards of glass and climbed through.

I had watched so much horror in a week. I'd seen maimed children on stretchers, people murdered in the street. I should not have been jolted by the sight of a kitchen, yet I was. We kept an orderly house, with everything in its place—and now it was in shambles.

For the first time it struck me that my life might never be the same.

The stove was wrenched from the wall and knocked onto its side. Pots and pans and dishes were strewn over the floor. The kitchen door had been looted. I picked my way like an earthquake survivor into the family room. What furniture was left had been dragged toward the door to the garage, the shortest route out. The television and VCR were gone, along with the sofa and Persian rug. Another TV, from the guest bedroom, must have been dropped in someone's haste—the screen was cracked. They had even tried to take up the wall-to-wall carpeting before quitting midway.

Broken glass crunched under my feet. My diplomas were on the floor next to treasured family photographs: Iman at her nursing institute. My kung fu championships. My sisters. Baby Abir. There was a pretty picture of my mother, ripped out of its wooden frame with silver inlay.

The frame was gone and the photograph smudged. It looked like it had been stepped on.

I went up the spiral staircase toward the second floor, where I hoped to get money and documents from our safe. Halfway there I could see our oak wardrobe, jutting out the door to the master bedroom. There was something sad about that bulky piece stuck like a fat man, neither in nor out.

A noise came from below. I went back down the stairs. In the doorway to the kitchen stood Abu Amar, with his son peering over his shoulder. All the men of their family looked the same: lantern jaws, big ears, brows and mustaches so bushy that they seemed of one piece.

"God brought you!" Abu Amar said, a wooden club in his hand. "Come with us respectfully. Neither God nor man can accept what you've done!"

I came off the last step and held my hands out, entreating them as I inched back toward the garage. "Abu Amar, we are neighbors," I said. "We should not fight."

Abu Amar shouted, "Stop, or I will break you!" He rushed me with his club held high.

In all my life, since my first lesson with Mr. Alaa, I had used kung fu in anger three times. The last was six months earlier, after a man harassed Iman at an orange-juice stand. When a situation demanded it, though, I did not need to stop and think. I came out of my crouch with the Slicer, a short, strong kick that draws its power from the lower leg. It caught my neighbor in his diaphragm and knocked him to the floor, gasping.

The son said, "I'm going to get my rifle and kill you!" He ran for the back door, screaming, "Mohammed is here!" I caught him in the kitchen with a jumping kick to the back of his neck: the Lynx. He bounced off the wall and fell unconscious against the toppled stove.

We had made enough noise to attract other bounty hunters; I could not linger. I went into the garage and found its door wide open to the street. Iman's car had been stripped of its tires—and where was mine? I ran around to our front door, also broken. In front of it lay Mareq, our four-year-old German shepherd, in a sticky red pool. He was a brave dog—gentle with our daughter, fierce with anyone who so much as raised his voice with me. Mareq would have given these intruders a fight.

There was more blood in the front garden, and I morbidly thought the worst. I imagined Iman or Abir or both of them lying there. I had never understood suicide, but for a few seconds it seemed reasonable. Even practical. When your life is *over,* you end it.

I looked across the street and there it was: my black Volvo sedan, parked by the house of Abu Khalil, the head of the local power company. With the driver's-side window smashed, it was simple to get in and hot-wire the car away, a trick I'd learned from my driving instructor. As I sped off, I caught a glimpse of Abu Khalil in my rearview mirror. He was standing in the street with nothing on but his boxer shorts and a look of astonishment.

As the car squealed into the roundabout, I saw three bodies lying by the curb. Neighbors were bringing blankets, but one man in a cloak and turban was still uncovered. I glimpsed a familiar gray beard—our mullah! He was the muezzin for the mosque a block from our house, the one who called the faithful to prayer five times a day. I myself went only a few times a year, when Baba was visiting and wanted company, but I respected this man. He was a good-hearted grandfather, devoted to the neighborhood. When my daughter was sick with the measles, he came and read her the Koran to ward off the evil eye. After that, Abir would bring Iman's home-cooked food to the mosque for him. She was so attached that I would tease her, "When you grow up, I'm going to marry you to the mullah." She became shy around him, and I felt bad—until I missed a fine cotton shirt, worn once. Abir had given it to the mullah.

When he came by to collect sweets for Ramadan, the mullah would complain about the regime and the "donations" we were forced to make to "charity." Had some passing comment reached the wrong ears? There was no way of telling, though our neighbors would sift through the rumors like archaeologists at a dig. Only one thing I knew for sure. Another good man was gone.

CHAPTER SIXTEEN

A RE YOU *CRAZY?*" Ahmed looked at me in disbelief. He and his wife and Hassan yelled at me all at once: "What are you doing here? Go now! Run away!"

But I was not leaving Baba's house without some answers. I asked them, "What happened?"

Ahmed said, "Are you kidding me? The fedayeen are looking for you. Do you know what fedayeen are?" When I asked about Iman and Abir, he said they were safe with Beijia, Iman's aunt—"but what about us?" He was angry but mostly terrified. It was not the time to share my plan about moving to the Marines' camp.

I would not feel secure until I saw Iman and Abir with my own eyes. The fedayeen were resourceful. It was not beyond them to track my family to Beijia's. I quickly washed my face and hair and changed into Hussen's clothes. I told Hassan to come with me and ran down the stairs. Baba was waiting at the front door, aggrieved by his wayward son: "Why did you put all of us in such danger?"

I told him briefly of the POW and added, "What happened, happened. I did this to save a life. Didn't you bring us up to help people? Haven't you risked your life for another?"

I could see his face softening. "All right," my father said. "If you do good, good will find you."

As we set off, Hassan told me what had happened to the Volvo, according to the scuttlebutt. Two thieves had come to strip my car, as they had Iman's. Outside our garage they had a dispute. One knifed the other, which explained the blood in my front garden. In the meantime, Abu Khalil had come from across the street. He told the thieves, "Don't touch this car, it's mine," and chased them away with a gun. Abu Khalil knew I was on the fedayeen's list, as good as dead, and so he claimed the car as his own. He broke the driver's side window, hot-wired the ignition, and moved it to his house. No wonder he'd been shocked to see me drive it away!

I had picked Hassan over Ahmed for this trip because he had no children to leave fatherless. It was a fortunate choice. At one Baath Party checkpoint, they ordered us out—they needed a civilian car for some urgent business. Hassan showed his police captain's ID and said we were on our way to a big meeting. (I tried to look suitably thuggish.) The Baathists relented, and we passed through.

Hassan was an easygoing man, but that day he kept muttering, "We're all going to die." His pessimism was infectious. It was after five o'clock when I knocked on Beijia's door. When it opened, I thought, I would know my family's destiny.

"*Mohammed!*" It was Ahlam, Beijia's daughter. She called out, "Iman, it's Mohammed!" and only then did I return to the living. Abir came running and jumped into my arms. Iman was awash in tears. I felt guilty, again, for leaving with no word.

I said, "When I went back to our house, the fedayeen came."

Beijia must have misheard me—she thought the men in black were down the street. Panicked, she ran toward her back door but tripped and fell. She was a heavy woman, and it was not easy to lift her. I said, "We'll have to call a crane." Beijia laughed; she would be all right now.

I told Iman, "We're leaving Abir here with Hassan. Go change into your uniform—we have to go to the hospital."

"What's going on?"

Impatient, I said, "I remarried!"

Iman looked at Hussen's outfit and laughed: "It doesn't look like you're dressed for that."

We traded stories as I drove under the darkening sky. When I told her about the mullah, Iman was aghast. The fedayeen had arrived at our house in four cars that morning, she said. Six men came to our door for me. When Iman told them she did not know where I was, they ordered her to come instead. But the mullah was there to intercede: "If you take her, you will never get Mohammed—he will run away. Leave her here with me. When Mohammed comes, I will bring the whole family to you."

That seemed reasonable to the fedayeen; a mullah's word, after all, could be trusted. As soon as they left, Iman took Abir out the kitchen door. They walked all the way to Baba's house. Ahmed drove them to Beijia's, a safer place, while Hassan went to try to warn me.

Iman was in tears. "You did a good thing," she said, "but you didn't think about us. You should have trusted me. I was so worried—I didn't know what happened to you. I thought they had killed you."

I said, "Iman, this is not the time for that. We have to get the information."

She said, "It is crazy for us to go to the hospital. But I've been your wife for seven years and I know how you are. If you want to do something, no one will stop you. So what do you want me to do?"

We were close now. "Okay," I said, "let me tell you what we need."

We parked by the nurses' entrance, where they checked IDs, and I held my breath as we approached. The guard knew Iman and waved her through, barely looking at me. Everything changed in a war, I thought. In normal times, the regime would have caught me within a day. But with the phones down and ten crises an hour, their intelligence was compromised. People like me slipped through their hands.

My plan was to make my survey of the floors while Iman got in to see the POW and learned her name and condition. If all went well, we could bundle my family off to the Marines the next day.

I'd thought the evening hours would be better for our work, with fewer people to notice us. In fact, it was the opposite. They took *more* precautions at night, posted more security. With visiting hours over, I was more conspicuous. Eyes bored into me up and down the corridor. I stuck to the ground floor and tried to look harmless until my wife returned.

"It's no good," Iman whispered. Fedayeen were swarming the second floor. When she went there on the elevator, a guard informed her that only floor staff could get off. Iman made up a story about Warda, a second-floor nurse—that Warda's father was a patient on Iman's floor, and he was suffering and needed to see his daughter. The guard told Iman that she'd have to call Warda on the house phone from downstairs. If he let her in, he said, he could be executed.

"Let's go outside," I said. There was no point in staying until someone pounced on us. I couldn't even be sure if the American was still in the cardiac unit. They might have moved her or—the thought tore at me—put an end to their complication. I would have to try again the next day.

On our way out, Iman talked her way into the hospital jail for a quick look. No POWs, she told me.

To use the time, we cruised around the walled perimeter, checking the odometer at points of interest. There were three gates into the compound, each with four or five guards carrying rocket-propelled grenades. I also counted seventeen late-model Toyota pickups, parked by various construction sites. Security men had camped in the half-finished buildings, thinking they'd be safer than in their nearby headquarters.

To keep the numbers straight, I devised a number system with my right hand, counting from the top of each finger down. I used my pinkie for military vehicles, with each joint counting for twenty. My ring finger stood for the number of armed men in and around the hospital. If I ran out of joints, I'd use the creases on my palm.

The compound was flanked on two sides by houses, on a third by office buildings. The only clear space for a helicopter was to the south, a vacant lot where children played soccer.

We finished our circuit at eight o'clock. As we left, I had a strong foreboding. What if the fedayeen had raided Baba's house and were now waiting at Beijia's? Better to lie low. I made a sharp right turn to the city center, toward the gym where I taught kung fu. The place was undisturbed; I had to break the lock with a crowbar. I wearily took the massage table for a bed, while Iman grabbed a mat on the floor.

I had forgotten that my gym was two hundred yards from a Jerusalem Brigade operations center. The blasts came nearer and louder. The power went out, leaving us in the glow of a charged fluorescent "candle." Our building shuddered and Iman cried out. A vase had fallen from a bureau on her head. It hurt her, and she said, "Did you bring us here to be killed?"

She was right. There was no safe place for us in Nasiriya that night. Better to stay on the move than to sit in one place—better to go to the Marines, where we had friends. I gathered some photographs and the videos of my title matches. "Let's *go*," Iman said. But there was one more thing. I opened my closet and picked out a green suit with a tan shirt and green tie.

Iman said, "So now you want to dress up for our funeral?"

No, I said. I wanted to look like a lawyer when I showed up at Saddam Hospital the next day. A professional. A person no one would suspect.

We stopped a distance from Beijia's and waited half an hour in the dark. I moved softly up the walk and put my ear to the front door. I heard Abir trilling an Iraqi children's song: *When are my father and mother coming?* Someone—Beijia or Ahlam—asked her to quiet down. All was well.

"Where are you going?" Beijia said, after answering my knock. To Baba's rental house across town, I lied. I took Abir and also Hassan, because my brother needed to see where he'd bring the family the next day.

There were eight checkpoints between us and the second Marine camp. I dreaded them all. I had needed a special import permit to get my Volvo—a year-old model—through the United Arab Emirates. Only two or three of these cars were registered in the entire province. A simple bulletin could undo us.

It was hard to see the holes in the road. We crept along, with plenty of time for hindsight. Now I saw that I'd been driven by impulse and a cloudy brain. Hassan's ID was no shield against the shelling that traced the sky like fireworks. We could have—*should* have—stayed at Beijia's. Corey had been clear: to go at night to the Marines meant likely death. And if we somehow got through, I would have to return to the city tomorrow, yet another trip through the war zone. How many could I make before my luck ran out?

At the eighth checkpoint, the guard said, "Sir, do you really know where you are going?"

Toward Islah, I said, the village straight ahead.

The guard said, "Just be careful that you don't take the right fork, because you'll end up with the Americans."

A mile and a half later, I took the right fork. It dawned on me that the Marines had no way to tell who we were. Unlike the regime, they knew nothing about the Volvo; I had not been sure which car I would be using, so I had not mentioned it. From the backseat I could hear Iman's shallow breathing. I could feel Abir's wordless fright. Had I missed my chance to save them?

I stopped and asked Hassan for his undershirt. I tied it to the aerial and rolled forward.

A flare streaked through the sky. It made no sound: the quiet before the storm. Once the parachute popped and it began its slow descent, the flare would illuminate the area like a movie set. They could hit us with ease, from the air or the ground.

I slammed on the brakes. Abir let out a yelp as she and her mother were hurled forward. I threw open the door and jumped out, my hands raised high. The yellow-green light was intense, washing out the colors of Hussen's outfit. My fear was like a fist in my throat. I stood there exposed for a lifetime.

No sign. I got back in the car and drove forward at a walking pace, flashing my headlights. After five hundred yards, I stopped at a barbed perimeter fence. Suddenly the Marines were all around us, twenty or more of them, pointing their rifles at our windshield.

"Don't move!" We hardly breathed. *"Don't leave the car!"* We stayed. They flashed lights in our faces, and one of them said, "They've got a kid in there with them!" The men did not lower their guns. They knew, as we did, that Iraqi children had been used by suicide bombers more than once.

I called out my broken window, "Can I talk to you?"

From his distance, the Marine said, "Talk, but don't move."

I said, "My name is Mohammed. I spoke with Corey." And I gave them Corey's password: "One-four."

The man in charge said, "Is that one-four or one-oh-four?"

"One-four."

"Are you sure?"

"Yes, I'm sure."

"Get out of the car—just you." The Marine scanned a piece of paper, which apparently held my description. He came up to me with a grin and said, "Welcome!" We were safe.

They took us to a tent, fifteen feet square. There was food and drink, and a stuffed animal for Abir. She accepted it with wide eyes, still too scared to speak. They brought pails of cool water for our feet, cots and blankets, and many assurances that "Saddam will fall, don't worry." In exchange, I wanted to give the information I had gathered outside the hospital. I wanted to accomplish *something*.

The one in charge said, "No, don't give it to us. It's not our job. You know what you're supposed to do."

I knew. I had to get back to Corey. This camp was a haven for my family, no more than that. My real work lay ahead of me.

I am not a worrier by nature. When the sun sets, my mistakes are behind me; the next morning brings a fresh start. But this night was different. I couldn't stop thinking about the day in front of me. Where would I be at the end of it?

And the POW—would this day bring her closer to her home? Or to her grave?

War shook the earth. I was too restless to sleep. It was still dark when I spoke to Corey by wireless telephone. He had two things to tell me:

First, there would be a break in the bombing of Victory Bridge between one and one-thirty that afternoon. I needed to cross in that window.

Second, I should not come to them after dark. He *meant* it this time.

When I said good-bye to my wife and daughter, Abir wrapped herself around my leg. I needed help from Iman, but she clung to me from the other side. A strapping Marine broke into tears. It seemed strange until I saw that he was only a teenager. Was he missing his mother, his sweetheart? Would he see them again?

"Please don't cry," I told him.

My legs were melting. I could call Corey and tell him I was sorry, but I had changed my mind—there was nothing more that I could do. I had made a good try; no one could blame me. I could go back inside the tent and sleep till noon, and stay here with my family. When the war ended, I could go home and die someday in my bed, an old man.

Or . . . I could return to the lion's den, a place where many wanted me dead. I could head straight into their arms, and it wasn't God who was bringing me, as Abu Amar had said. It was something I could not quite explain, even to myself.

These Marines had called me a *hero,* and I must admit I

liked the sound of that. But heroes made sacrifices. Heroes paid whatever price. I had been through a lot, yes, but I had not finished the job.

"Iman," I said, more sternly than I meant to, "if you don't take Abir and get inside the tent, we will all be sorry for a long time."

I nodded to Hassan and we got in the car, heading north into Nasiriya.

CHAPTER SEVENTEEN

I MADE MYSELF A PROMISE. From here on I would take my time and plan each move ahead. I would cage my feelings and work with a clear mind. For this one day, I would be as cool and ruthless as the fedayeen.

I turned off near our hotel, a good place to drop Hassan, get a shower and shave, and change into my green suit.

My brother said, "You packed it, then?"

I said, "It's in my gym bag, in the trunk."

And he said, "But the Marines unloaded it from the car when we got there." I had not noticed, and no one had thought to tell me—a gym bag seemed trivial in the scheme of things. Now I would have to wear Hussen's clothes into the hospital instead.

I decided to skip the shower and pulled over near Baba's house. It was my last good-bye, and I knew it would be a hard one. Law enforcement was not my baby brother's calling, you see. In college he had studied linguistics, and he might have been a teacher had not Uncle Ali Khayoun prodded him into police work. Hassan was built like a bear, but he had al-Hbeiba's tender heart. As he left me, he said, "I did not cry in front of my wife, but now I will cry, because one of us will die. I will never see you again."

I turned up the volume on my tape player—a sad song called "Hawalay"—and shouted, "I can't hear you!" Hassan chuckled through his tears. I said, "What do you think is so funny?"

He said, "I remember what Baba used to say, that you're always doing something unexpected. You always go your own way, by yourself." It was true. Whenever our family traveled by train, Baba would sit on one end of the bench, al-Hbeiba on the other, and all the children in between—except for me. Even as a young child, I wanted my own seat.

I loved my brother, but time was short. I told him, "Either you leave the car or I'll leave." Hassan left, with a pass for seven people into the Marine camp taped under his arm.

With the streets blocked by fallen trees and abandoned cars, I drove at a crawl. By the time I arrived at the hospital compound, it was after nine o'clock. I made two painstaking trips around the perimeter to confirm the numbers from the night before. I parked by the front entrance, in a space marked for an ambulance. If I had to leave in a hurry, the car would be handy.

I told the guard I was there for Dr. Hamida. I walked into the compound, toward the north hospital entrance and the elevators. When I saw some men in black on their way out, I circled back to the south, to the emergency door. A few people from my neighborhood were helping an injured man inside. There was too much money on my head to trust them. I wheeled away and turned again to the north.

On my way I met a worker I knew, the morgue attendant, en route to the hospital shop for a pack of cigarettes. He was an older man, cheaply dressed—except for his

well-crafted, khaki-colored boots. U.S. military boots. I
went with him to the shop and said, "Let me get these cigarettes for you. Hey, those are beautiful boots. Where did
you find them?"

The man said, "I got them from one of the dead Americans—maybe you're disgusted with me."

I said, "Not at all, I would like a pair myself. How many
Americans did they bring in?"

The attendant counted aloud: "Seven . . . eight . . . *nine*.
But there's only two left in the morgue. We got too
crowded in there, so we buried the rest outside." When I
asked the location, he became evasive. "Do you want the
boots or not?" I told him I would bring the cash the next day.

It was close to eleven o'clock. Allowing fifteen minutes
to drive from the hospital to Victory Bridge, and five minutes more to cross it, I had just over two hours to do my
work and avoid the shelling. I would want no more time
than that, I thought. Every second inside was a chance to
be caught.

The key to my plan was Hamida. She had the authority
to enter the cardiac unit—to see the girl herself if the Fu
Manchu guard manned the door, otherwise to try to bring
me with her. Then she could join Hassan and the rest for
their trip to safety.

When I made this proposal in her office, my sister-in-law
said, "This is very dangerous, Mohammed—for you and for
me." She was not even sure the POW was still in the same
place.

Let's at least find out that much, I said. Eyeing me
unhappily, Hamida rose from her chair. We took the stairs
to the second floor, where I could wait inside the doorway
while she checked the guards and waved me on or not. As
she walked to the cardiac unit, I peeked out after her. Far-

ther down the corridor, I made out a small crowd of people and a flash of Baath Party green.

Hamida returned to me frowning. It was a bad time, she said. The Fu Manchu was not on duty, but they'd brought a wounded Baath official to the floor. He held a high rank, and several soldiers and fedayeen were there with the family.

And she said, "I think they might have moved your POW to another room. It was dark in there through the crack in the door." My heart sank—was I too late? As we trudged back up to the fourth floor, Hamida tried to convince me to give up. The other day, she said, she happened to be in the cardiac unit while someone from security interrogated the prisoner. At one point he pulled the girl up by her hair.

Hamida said, "If they are treating the American this way, how will they treat us if we try to help her? Forget about her—she is finished. One of her legs is so bad that they're going to amputate tomorrow."

I was crushed—how did she know? Hamida's friend, a doctor I will call Dara, had mentioned that her boyfriend was set to perform the operation. Hospital officials thought it was too much trouble to do all the surgeries the POW required. An amputation was less bothersome. And if she died, she died.

This was terrible news. Conditions at Saddam Hospital were so poor that any surgery was high risk. To make matters worse, the recovery rooms were scattered on various floors. If the amputation went forward, the Americans couldn't know where to get the girl.

I said, "Hamida, you have to stop this."

She said, "It has nothing to do with me."

And I said, "Can't you do something to get it put off a few days? *Please,* Hamida."

She thought a moment. "I'm covering for Dara tonight as a favor. I could ask her to talk to her boyfriend. She says he'll do anything she asks when they go to bed."

May Dara be persuasive . . . I asked Hamida to wait for me in her office for half an hour, when we'd try the second floor one more time. Then she could go home.

I glanced at my watch: half past eleven. To stay ahead of the shelling, I'd need to be in my car in ninety-five minutes. I took the elevator to the sixth floor and climbed a long flight of stairs. The door to the rooftop was locked. If I kicked it open, the noise would expose me. I descended the stairs and found the man I needed: Abu Ibrahim, the floor janitor. I told him I had helped to bring a dead man into the hospital. Iman had washed my clothes, I said, and I needed to hang them on the roof to dry. Did he have a key?

The janitor said, "I don't have that key. Look, I'll try to help you, but I need to ask something from you as well." Abu Ibrahim was married to a much younger woman. He wanted pills to boost his virility, but the drug was expensive and given only by prescription. Perhaps Dr. Hamida could find a way—

"No problem, I'll get you some," I said.

The happy janitor went off to find a key. It took him ten minutes to return: an unforeseen delay. As he fiddled with the lock, he said, "When can you get me the pills?"

"Tomorrow," I said. He opened the door and bid me good-bye. As I stepped onto the concrete rooftop, I felt better, easier. This was my first success. I could see that there was unobstructed room for a helicopter and—most important—no sign of explosives. The Marines could land here if they wanted to.

Back on the sixth floor, I began my survey of the interior. Each floor of Saddam Hospital was built around one

long corridor, running north to south. Most had twelve
rooms on each side; each room had from one to six beds.
At the corridor's midpoint there was a station for doctors
and nurses, a large wall closet for supplies, the main stairway, and the elevators.

I had to inspect each room for the girl or any other
POWs who might have landed there. I looped the corridor,
checking only to my right; it would seem less obvious that
way. I looked into doors that were open, knocked and
opened any that were shut.

I saw nothing unusual at the start—only injured Iraqis,
with few nurses in sight. There was no new security equipment, no cameras or police dogs. Three Baathists ate their
lunch in the nurses' room, their wood-butted Kalashnikovs
close at hand. I quietly passed by, adding the number of
men to my ring finger, their weapons to the account on
my middle finger.

A dozen steps later, four fedayeen—two in plainclothes,
two in ninja black—came into view. I darted through the
nearest doorway, into the burn unit, to wait for them to
pass. The unit had ten beds, and the smell of seared flesh
was overpowering. Next to me were two corpses, their
blackened feet poking out from under their sheets.

Sickened, I lurched back into the corridor. There was
another bed, part of the unit's overflow: a woman screaming in agony, her husband and two crying little boys at her
side. The victim's face was a twisted mask, her hair matted
into spikes. A nurse treated her to no effect. Morphine had
been cut off to Iraq. Other pain medications—along with
clean sheets and blankets—were reserved for the well connected. The rest, like this one, could only suffer.

I finished my round in fifteen minutes, behind schedule.
To stay invisible, I took an out-of-the-way emergency stair-

case at the north end of the corridor, by the bathrooms. My tour went faster on the next three floors, five minutes for each. There were no gunmen to hide from—and no sign of the POW. My confidence grew that she had yet to be moved.

It was twelve-fifteen. I had less than an hour left.

Saving the cardiac unit for Hamida, I passed to the ground floor. Here my work grew more challenging. The military's staging center was housed in the hospital director's office. Its windows were draped over. A stream of regime types, mostly fedayeen, flowed in and out the door. To count them properly, I found a spot in the lobby, fifteen yards away. When any headed toward me, I bent to tie my shoes. Or I collared a nurse or orderly and asked about the first name that popped into my head: "Have you seen a woman named Batul admitted here?"

Within half an hour, I tallied a general and two colonels, the local Baath Party leader, and the director of intelligence. I passed the second joint of my ring finger: a total of forty-one armed men. When no one new came out for five minutes, I moved on.

It was twelve forty-five.

In the basement I found a supply station and a place where the staff stored their food. A third room's door was unmarked, but had many locks. When a friend of Iman's came to get her lunch, I said, "What's going on with all these locks? Is that where you keep your salaries?"

She said, "No, there are explosives in there."

At twelve-fifty, I huffed up the four flights to Hamida's office. Her door was shut for an exam. My window had shrunk to fifteen minutes. I wanted to charge in and steal Hamida away, but that would create too big a stir. I stood outside and scowled at my watch, trying to will the hands to freeze. I saw a man smoking in the stairway door and went to say hello. He offered me a cigarette and I took it; though I had never smoked before, I was desperate to kill some time. I coughed and stubbed it out.

The patient left at one o'clock. I ran into the office and grabbed Hamida's hand: "I have to see the POW *now*."

"What about the guards?"

I laid out the scheme I'd hatched. I would fake heart palpitations, and she would bring me into the unit for an EKG. A fine idea, Hamida said, "but they won't let us in without the paperwork."

Back to the ground floor. A conversation between Hamida and a talkative clerk. A leisurely hunt for the proper forms. At one-ten, another conversation. A 750-dinar fee. The pink and green papers presented and filled out.

We returned to the emergency staircase, the one passage left unguarded by the fedayeen. When we reached the second floor, I looked at my watch: twenty past one. My window had closed. I didn't bother checking after that.

From the far end of the floor came a high-pitched keening. The Baath official had died. The men in uniform had left the family to its grief; the corridor was clear. There were only the two fedayeen guards at the entrance to the cardiac unit—the only way in and only way out.

Hamida moved forward with a doctor's purposeful stride. I followed a step behind, stiff with fear, my heart pounding fast and hard. I thought to myself, *I might not have to fake this.*

I had seen these guards in the hospital the week before, but we'd never spoken. One stood as he read aloud to the other from a newspaper. He was a bodybuilder, with a chest and shoulders that almost burst through his yellow T-shirt. The other fedayee was seated, a man of normal build in black slacks, white dress shirt, and striped black-and-white tie.

"Hi, doctor," the reader said.

Hamida said, "How are you? I have a checkup paper for this patient."

The bodybuilder skimmed the paper. "Sorry, Doctor, but we have strict instructions. Didn't you know that no one can get inside here except cardiac staff?"

Hamida said, "Yes, but this is my brother."

The seated one, who seemed to have the last word, said, "Oh . . . all right." He told his partner: "Take care of her, so she'll take care of us." With medicines and first-aid supplies so scarce, everyone needed a friend in a hospital.

The bodybuilder gave us an oily smile and opened the door. As we stepped into the outer chamber, my heart truly was palpitating. I was so near to my goal.

"Well, Doctor, what can we do for you?" I nearly jumped. It was Abde Hmoud, the unit supervisor, a small and pompous male nurse.

Hamida held up my paperwork. "It's my brother. I need to give him an EKG." As she spoke, I craned my neck to look into the rear room on the right. Like the rest of the unit, it was lit only by the sun. But I was positive that I could see a small form under the blanket . . .

Abde Hmoud said, "Let me do the EKG for you, Doctor." He might have insisted if his girlfriend, a nurse named Leyla, hadn't come in at that moment. She was on break and wanted company.

"All right," the supervisor said. "Doctor, I'll be right back if you need something."

We were alone in the chamber. In a low voice I said, "Hamida, let's go and see the POW."

"What if somebody comes in?"

"No, no, we'll be all right if we hurry."

Knowing she could not stop me, she led the way. After so many close calls and near disasters, it seemed almost too easy. The turn of a knob, the swing of a door . . . Hamida stopped at the foot of the bed and immersed herself in the patient's chart. I walked on by and into the small room, to this girl with the bandaged head. My mind counted my steps, filing one more bit of data for when it might be needed.

But my heart was another thing. My heart was full—*I had found her!* The POW lay flat on her back, breathing evenly, her eyes tracking mine. A white blanket covered her to her neck, except for her right arm, which was heavily bandaged from wrist to shoulder. A tube snaked underneath to the other arm, feeding some clear liquid.

The prisoner's face was bruised on both sides, with a swelling above her left brow. Every few seconds she winced in pain. Otherwise she was silent, watching. In the soft light, she looked even younger than before.

"Both legs are broken," Hamida said without looking up, her fingers riffling through the pages of the chart.

I felt a surge of pity. I wanted to carry this poor girl out of Saddam Hospital, then and there.

"But she can be moved," Hamida went on. "The IV is only for nutrition."

I nodded and noticed some writing on the cover of the patient's chart. It was the POW's name in Arabic: *Jessica Ameriqia.* "The American Jessica."

At last I knew her name.

I needed to tell her something, but my English made it difficult. I tilted my head and offered the one greeting I could remember: "Good morning, Jessica."

And she said, "Good morning, Doctor." (What else would I be, after all?) In spite of everything, she gave me a small smile. That is when I knew she was very brave.

I ached to tell her everything. That she shouldn't worry, because I would help her. That her American friends would not leave her behind. But I didn't have the words, and I couldn't have used them, anyway. Corey's instructions were clear. My assignment had to stay a secret, from *everyone*.

I could only tell her this: "Don't worry." She seemed puzzled, as though she had not understood, so I repeated as clearly as I could: *"Don't worry."*

The girl looked up at me and blinked back a tear.

"Let's go," Hamida said. "We don't want anyone finding us here."

We'd been in the room a minute, maybe two. *You won't die here*, I thought as I reluctantly left the bedside. *You are not dying anytime soon.* I knew this as I knew my name: Jessica would survive. Her condition was stable. I had all the information. There was but one small step remaining.

They only needed to come and get her.

CHAPTER EIGHTEEN

M Y SISTER-IN-LAW was prophetic. Just seconds after we'd softly closed the door to Jessica's room, even before I could lie down by the EKG machine, Abde Hmoud bustled in to tell Hamida that she was needed on the fourth floor. I made a show of tucking in my shirt, as if we'd just finished the test.

When Hamida thanked the nurse and turned to go, he said, "But you did not get your report." Now she was flustered; she wasn't sure what to do. "It's okay," he said. "You go. I'll give it to your patient." With a nervous glance in my direction, she left.

When Abde Hmoud checked the machine, he was baffled: "What's this? Funny, I used it the other day and there was nothing wrong. Dr. Hamida knows how to use this equipment. Why is there no report?" He was determined to solve this mystery. "I know—you must have moved. That's why it didn't register."

I said, "Yes, that must have been it."

Abde Hmoud told me that I would have to repeat the test. *Better to keep him happy,* I thought as I lay on the bed.

When it was over, he looked at the results and said, "You're in better shape than I am!"

Armed with my bill of health, I moved to the outer door, preoccupied by the trip ahead of me. It was nearly two o'clock; Victory Bridge would be a dangerous place by now. It took me a moment to notice that they'd changed guards. The new man in the chair, engrossed by a hand-held computer game, was the sleeper of three days before. As I pivoted into the corridor, there was the Fu Manchu, ten steps away, returning from the men's room. He rubbed his hands with a paper towel, then his face—and then, his eyes alight, he flung the towel down.

"You, *again?*" he roared. "What are you doing? Come here!" He snapped at his partner: "Why did you let him in there?"

By then he was upon me, his meaty hand on my shoulder. His fingers were like steel rods. He spun me around to push my face against the wall, slipping his gun from his belt at the same time. But before he could grip my hands to cuff me, I swung my right arm back from the elbow. My fist bashed the point of the Fu Manchu's nose, and I heard the cartilage crack. Another backhand blow to his face, a chop and kick to his stomach, and the hulking fedayee was stretched out woozily on the linoleum. His gun clattered to the floor.

The sleeper woke up: "Huh? *Huh?*"

I wanted to go for the Fu Manchu's Series 16—if people saw me running with that kind of weapon, they might think that *I* was the fedayee and get out of my way. But the sleeper was fumbling with his own gun. He might beat me to the draw; time was not on my side. I turned and ran.

As I reached the stairway, the sleeper stooped to help his partner, who furiously brushed him away. Audibly in pain, the Fu Manchu yelled, "Call them! *Call* them!"

I vaulted down the stairwell and burst through to the ground floor, wailing, "He's gone! My brother is gone! God have mercy on him!"

I heard someone saying, "What's going on?" As I neared the front entrance, there was a thunder of boots and the cry: "Arrest him!" I had a twenty-yard lead and a better parking space—enough, I hoped, to see me through. As I revved toward the main gate, past openmouthed nurses and orderlies, I heard the stutter of automatic weapons. My car wasn't hit, and none of the pedestrians went down. *They must be firing in the air,* I thought.

One vehicle at a time could pass through the gate. I beat an incoming taxi to the spot; the cab jerked into reverse. Blasting my horn, I swerved right onto a narrow two-lane street, pocked with holes. I drove through them, bouncing till my teeth rattled. Anxious that the car might stall, I slowed to forty-five miles an hour. Soon I heard a pinging of metal on metal. They weren't firing in the air anymore.

I made it to the four-lane Guitar Street and shot southwest to Victory Bridge. With no traffic lights, no traffic, and room to maneuver around the larger holes, I sped up to seventy miles an hour. There was still shooting, but no more pinging—I must be putting more distance between us. I stole a look over my shoulder; a black Land Cruiser was fifty yards behind me. The Fu Manchu hung out the passenger window with a Kalashnikov, pumping out six shots a second. I could see the red flicker from the barrel as they jounced along.

The closer to the river, the heavier the bombing had been. At the bridge approach there were more craters than road, gaping pits as much as ten feet wide and six feet deep. I hurdled the sidewalk around a burned tank and eased to twenty-five miles an hour. The Land Cruiser closed ground with its four-wheel drive. I could feel the Fu Manchu's breath on my neck.

Victory Bridge spans sixty yards over the Euphrates, with another thirty yards on either end over land. Just before I got on, a bomb exploded in the water with a tremendous splash. I gunned into the bridge's one open lane, on my right. If someone came the other way, the game was over— I'd be trapped. Another bomb struck a building just across the river, a former children's arcade that now stored tanks and weaponry. I saw a Cobra hovering lazily over the river, ready for anyone foolish enough to cross. Would the fedayeen see it, too? Would they call off their chase?

I was ten yards or so over the water when the next bomb landed, the one that changed the equation. So much happened in a split second, distorted into slow motion: a deafening boom and a wave of heat on my face . . . the road heaving as I slammed on the brakes . . . the car pulling wildly to the right, smashing into the walkway's curb . . . the air bag, like a punch in the face . . . the smell of gunpowder . . . a final, screeching collision. I thought I would pitch into the river. The bridge railing buckled, but held.

For a long moment I sat there blinded by the air bag, coughing in the thick black smoke. I was seized by the worst pain of my life; my head was on fire. The worst of it was my left eye. Someone was twisting a knife there and would not stop.

I brushed the bag from my face, but I still could not see. I brought my hands to my face and felt the wetness. There must have been blood all over—was I hemorrhaging? My tongue found another wound inside my upper lip, a deep gouge from my teeth.

I looked down to free my seat belt, but couldn't find the latch. My right eye was blurry and I couldn't open the left one. I couldn't even be sure it was still there.

I had to locate the fedayeen. Though the Fu Manchu hated me enough to shoot me on sight, the others would want to use torture. I could not let them take me, but I was too hurt to fight and too sapped of strength to run. If I had no other choice, I would jump into the river and let the current take me where it would.

Slowly I got out of the car. I stood dazed in a cloud of black smoke and white dust, and then I lost my balance and fell. When the hammering pain let up, I stood again. I faltered into the roadway and looked back. The fedayeen had vanished. There was no one on the bridge.

Back at the mangled rail, I looked down toward the riverbank—and there was the Land Cruiser, fifteen feet beneath me, its nose a few yards from the water. Judging from their skid marks, the fedayeen had lost control on the bridge approach, where there was no barrier. The vehicle's roof was caved in. A man sprawled facedown on the hood, halfway through the shattered windshield. He wasn't moving.

The bomb must have fallen closer to them than to me. Was that the difference between life and death—a few yards, a fraction of a second?

A new explosion turned me around. They had hit the arcade again, sending it up in flames. There was no sense in delaying, and no question of using the Volvo. The front end was smashed into the tires. I ran toward the burning arcade, weaving from road to walkway around the craters. I needed to cross as quickly as I could. As far as I knew, the regime still controlled the far bank. I was fair game for their snipers.

I forced my left eye open. It made out a hazy landscape, but the view gradually darkened. Something was very wrong. My head throbbed with each step. My chest was bruised from the air bag, and I stumbled more than once. But to quit was to die, so I ran.

Across the bridge, beyond the arcade, there was a walled

public garden. For protection I hugged the inside of the wall, moving east, parallel to the river. The garden was in bloom: yellow sunflowers, white princess of the river, aromatic eucalyptus.

I turned east through another empty neighborhood, into a narrow alley—a wrong turn, I realized, when I came upon the three men with the Dome of the Rock on their chests. I had seen the Jerusalem Brigade at work; they would call you over to check your national ID card, then let you go. But these three trained their guns on me without a word. As they tied my hands and fastened my blindfold, I said, "What did I do? I am not a soldier—I am a civilian."

"Don't speak."

Did they know me? Did they know what I had done? I was marched a short distance into some building. Though I heard my captors leave, I was not alone. A lively discussion ensued, with several different voices.

One man said, "Those guards are dogs, I tell you!"

Another said, "Do not curse them like that—you are traitors!"

And a third said, "Whatever you do, even if they put pressure on you, don't admit that we are soldiers."

I did not like the sound of this, especially since I'd been relieved of my ID card in the alleyway. In wartime, any man of draft age found without an ID was presumed to be a deserter. Most often they were stood against a wall and shot.

I heard a door creak open. My blindfold came off, and I opened my one good eye to a bad dream. I was in a garage with two armed guards and four other prisoners. One was my twin, with only one eye; another was a hunchback; a third was short and scrawny, barely a hundred pounds; the fourth was older, around forty, and as big as a tree.

The guards pulled the hunchback aside for a conference—and untied him and saw him out. I overheard enough to get the gist of things. These mercenaries had corralled four deserters and notified their superiors. No doubt smelling a bounty, they in turn had called in a death squad, the specialists. In the meantime, however, the mercenaries had learned that the hunchback was a friend's brother-in-law. They wanted to let him go, but needed a replacement to keep their numbers straight.

That was where I came in.

I went to the guard at the door and said, "I want to talk to the person who took my ID."

He looked at me blankly. "Nobody took anything from you."

We heard a car pulling in, and a new voice: "Bring them out here so we can see them!"

My cellmates and I were trooped out squinting into the sun. Our executioners lounged beside their white Land Cruiser, four surly-looking goons in European suits and open white or black shirts. Handpicked by Chemical Ali, these men were known as Saddamirs, after the way they wore their red headdresses, tied up at the sides rather than draped down. On ceremonial occasions Saddam wore his the same way, to show that he was always ready to fight.

"We can't do it here," a Saddamir said. "Bring them out to the field and we'll meet you there."

I was herded with the others into the back of a pickup, two facing two. The man who'd taken my ID was our driver. Another mercenary sat above us, a rifle across his knee, watching closely. They had bound us with socks, in a knot I probably could have unraveled. But my hands were in the guard's plain sight. There was nothing I could do.

This was too cruel, I thought. I had outwitted the regime coming and going. I had escaped the fedayeen and a likely death on Victory Bridge. And now I had been swept up at random, for something I had not done.

We were rolling down a long hill when the pickup slowed to a stop in a spinning of wheels. We were stuck in the mud. As the driver shifted though the gears, we sank deeper. Leaving his rifle on his seat, he pulled out a shovel to dig under the tires. His partner hopped out of the back to offer advice. It was the small opening I had hoped for.

"Look," I whispered. "We can die here or in the field, so let's do something."

The scrawny man was game. The giant was nervous, but willing to try. The one-eyed man—the one who'd cursed the guards—was so nervous that he couldn't move. We would go ahead without him. I told the others how to untie their hands, but warned them to keep them behind their backs until the time was right.

We could hear the driver getting frustrated in the mud. His partner said, "Why don't we get these four to do it for us?" Then he called up to us: "You men come down here and let me see what you can do."

The scrawny man winked at me and said, "But our hands are tied."

The guard said, "All right, I'll untie one of you, and he can do the rest. But no funny business." Nodding to the giant, he unlocked the gate and pulled it down. After the giant descended, the guard squeezed his rifle under his arm to loosen the sock—and that is when we jumped him. He had no stomach for a fight; he dropped his gun and ran down the hill.

As the driver tried to get up from under the truck, I tackled him. By that time the giant had untied the one-eyed man, and they, too, had run off. But the odds remained in my favor. The scrawny man aimed his new rifle at the driver's head. We all knew he would not mind using it.

"Where is my ID?" I demanded.

The driver was trembling: "Please don't kill me!"

I said, "We won't kill you, but I need that ID."

The driver handed me his wallet—with my ID, his ID, pictures of his girlfriends, a few thousand dinars. He said, "Take it all—take the truck, anything you want. Just let me run away and don't shoot me!"

Then we heard it: the chug of an engine in low gear. An Iraqi army jeep was hauling a water tank down the hill. It would reach us in a minute, at most.

The driver screamed, "Come! *Come!*"

The scrawny man was brave but he'd had enough, and ran off with the rifle. The driver broke my grip and went for his gun. I tackled him again; we held each other in a death grip. This mercenary had recovered his courage. He insulted me with vile language and said, "Today is your day to die!"

The jeep was pulling over; I had to act fast. I kneed the mercenary in his most sensitive spot. As he collapsed, retching, the soldier was on me. I landed a right elbow to his chin, and tried to grab his shirt to deliver another knee. I felt him spin away—and found myself holding a grenade I'd unpinned from his belt.

The soldier stared at me in horror and tore off down the hill. I blindly chucked the grenade and scrambled behind the pickup. . . .

When I came back out, the jeep was destroyed, the mercenary crumpled next to it. It was an outrageous stroke of luck—like something out of *Total Recall,* when Arnold repelled the Martian mutants. I had won, yet I felt only loss. In the moment I tossed the grenade, just before diving for cover, I'd glanced back over my left shoulder to see where it was going.

But I saw nothing.

My left eye was gone.

CHAPTER NINETEEN

THE AMERICANS had swept a mile or so to the west since my last trip, which meant they were that much closer. I reached the edge of the last neighborhood. Immediately to the east lay one more open area, a grove of date palms. Beyond that, the Marines.

The bombing was intense again. I had a feel for this ground by now; I guessed I had a mile to go. I passed a dead horse, and more scorched and battered tanks. Their steel plates were cool to the touch; Saddam's troops had retreated some time ago. Still, I knew that I could not relax. The Iraqi army was known for mining areas that it could not hold. Commonly they used children for the task, nine- and ten-year-olds who slipped through the lines. I held my breath and tried to walk on my toes.

By the time I heard the Americans calling out to one another, it was dark. For safety I dropped to my knees and crawled through the rocky sand. I stopped behind a thick palm on a small rise. Hussen's outfit was no help to me now; they would not see the colors. I remembered Corey's words: *Whatever you do, don't come at night.* I considered waiting there till dawn—until the sand danced around me from a round of automatic weapons. A searchlight panned the area, back and forth: the Marines' nightfall routine.

I had an inspiration. When I had changed at Baba's house that morning, I'd borrowed underwear from Ahmed. Normally I wore black briefs, but he owned only white ones. I used to tease him about it, but now I had my white flag.

I stripped down to the briefs and my sneakers. No shirt meant no bomb. Shivering in the cool air, I awaited a pause in the firing. I stepped out in front of the tree, my hands to the stars. The searchlight panned to me, passed by, swung back. The glare blinded me, and for a moment I felt self-conscious—*I must look ridiculous,* I thought. Then the guns resumed and I screamed, "I am Mohammed! My name is Mohammed!"

The shooting stopped.

"Come," said a voice behind the light. "Hurry!"

I was cold and afraid and in great pain. But when I bolted those last thirty steps of my mission, I felt only gladness: *I made it!*

As four Marines threw a blanket over me, one of them said, "Your shorts saved your life!" Everyone laughed.

Did I want them to carry me, or could I walk?

"No," I said, "I can walk."

How did the job go? they asked. Is everything okay?

I said, "Everything is good."

The lean-to was nearby. Corey was waiting out in front. "Welcome, hero," he said.

They gave me orange juice and another chicken soup MRE. This time I devoured it. A shower would have to wait, but they brought water for my face and hands and feet, and clean Marine fatigues. A medic examined me. I was fine, he

said. I just needed rest. My eye would require an ophthal-
mologist, which they didn't have in the field.

Corey said that I should sleep, but there were two things
he needed to ask me. Was the POW still in the hospital?

Yes, she was, I said.

And did I know her name?

I said, "Her name is Jessica."

All the men broke out clapping and cheering: "Lynch!
Yes! *Yes!*"

It was ten o'clock when they left me to my cot, but
there was too much noise to sleep. After half an hour I
gave up and called for Corey. Soon I was seated at a desk
with a map of Nasiriya, and a lamp with a jacket thrown
over it to foil any Iraqi snipers. Three Marine sergeants and
their aides clustered around me, firing one question after
another. Corey focused on the POW and her condition, a
second man on the hospital, a third on other buildings in
the area.

I told them everything I had seen, from the rooftop space
for a helicopter to the unmarked room in the basement.
They brought me graph paper and I made my first seven
maps. Corey sketched as I recalled the hospital perimeter
and surroundings, the outbuildings inside the compound,
the exact spot near the morgue where—Hamida had told
me—the Americans were buried. We used four different
felt-tipped pens: yellow for patient rooms; silver for the stairways; black for the route I took; red for points of danger,
where the fedayeen were posted.

My head was pounding. It was hard to focus on my fingers for the numbers they needed, or to work on the maps
with one good eye. The Marines asked the same questions
three and four times, to make sure of my answers. It took
all of my strength to keep up with them.

The mortars picked up again, from both sides. We

needed to move from the front, Corey said—back to the schoolhouse, in three jeeps. They took me into a different classroom, with the students' desks and chairs piled to one side. In the corner sat a bed, a real one. This was where the commanding officer slept, they told me, but for now it was mine.

There was a bathroom in the school, but no running water. Determined to have a shower, I got an idea. The Marines filled a large plastic bag with cold water, punched some holes, and hung it from the nozzle in a shower stall. I soaped away the grime and dried blood—it was a miracle to be clean. I washed my hair and found fragments of metal and glass. As they swirled down the drain, I felt secure. I was in a different place now.

After I got out, they placed a call to my wife at the other camp. She was crying: "Thank God you came back safe!"

For reasons of security, there wasn't much we could say. Abir came on the phone and said, "Papa, bring me a doll." Iman had told her that I'd been on a business trip, and my daughter expected her reward. I assured her I would do it, though I had no idea how.

Iman came back to wish me good night: "I love you, Mohammed."

"I love you, Iman." We had said enough.

One of the sergeants brought me a pill to help me sleep. As I drifted off, I felt filled with gratitude—for my new friends and good fortune, but most of all for *life*.

A landing helicopter roused me a few hours later, groggy from the sleeping pill. It was light out: a new day. Corey made sure I had hot coffee with my breakfast, and then we took another ride, to a technical institute building to the south. After escorting me down a corridor to a well-guarded door, Corey extended his hand and said, "I might not see you again. I have another job." He passed a document to a master sergeant named Jim, and he was gone.

My new team's headquarters was housed in the institute's gymnasium. There were six long tables with laptop computers, an overhead projector, and what looked like a telephone switchboard. Nasiriya street maps lined the gym's windows. Other city maps—Basra, Mosul, Tikrit, Al Kut—papered the middle of the floor.

I sat down with eight people, including Khalifa and a second translator, a Lebanese woman named Lara. I began by correcting some street names on the Nasiriya maps. Then I answered more questions about Saddam Hospital, in greater detail than before. The Marines seemed surprised by the number of armed men I had seen inside. When I told them that a Japanese company had built the hospital, they went to the Internet and found the types of windows the company used. They were leaving nothing to chance.

Whenever I provided something useful, they raised their cups in a toast. The rest of us drank orange juice, but Master Sergeant Jim toasted me with cold coffee. *He must be tough*, I thought. He was a medium-size man with a dark crew cut and a confident air—a take-charge guy, as the Americans would say.

They gave me aspirin every two hours but nothing stronger that might cloud my head. When I felt disoriented after lunch, they took me to the roof for some air and gave me binoculars. To the east I could see the burned-out hulks of Jessica's convoy. To the west, a hundred yards from us, there were two civilians lying in the sand and a man in a brown *jalabia* walking toward them. Lara said he had come to organize a proper burial. As we watched, however, the Iraqi passed the two bodies and kept coming our way. He was promptly arrested and taken to the classroom I'd slept in the night before.

Could I identify this man? He was about thirty years old, of average size, nondescript in every way. Something about him seemed familiar, but too much of his face was covered for me to be sure. They moved him to another room, removed the blindfold, and had me look through a hole in the door.

Then I knew. It was Qattaa Rouess, the Head Cutter—the most brutal fedayee in all of Nasiriya, a man infamous throughout Iraq.

There was a story behind his name. It was common knowledge that Uday Hussein had lost his manhood since an assassination attempt in 1996 left a bullet lodged near his spine. He would have poor young women procured—virtually kidnapped—for him and brought to his palace. When he was frustrated, he was apt to fly into a rage and have them killed.

A few of these women managed to flee Uday's palace and escape into Jordan, but the Jordanian government returned them. One of them, named Farah, was charged with prostitution, a capital crime under the regime. The Head Cutter was assigned to her execution. After beheading her, he wiped his sword and then sank his teeth into the bloody head. They actually announced this atrocity on Iraqi state television, adding a verse of the Koran to justify it.

The Head Cutter had entered the Marines' compound unarmed. His ID was from a soccer club. The Americans had no evidence against him, only my account. The man wanted to return in two hours with burial clothes—what should they do?

When I saw Jim hesitate, I suggested that they let the man go and see what he was up to. "But if he's coming back," I said, "be careful of him."

Less than two hours later, sentries confronted the Head Cutter before he'd reached their meeting point. This time he wore an *abaya*, which they ordered him to open. When he answered with a burst from his rifle, they killed him. In checking the body, they found several concealed grenades.

The Marines were happy with me that afternoon. I received a full-time bodyguard and the run of the school; I could enter the control center without clearance. When Jim pinned some sergeant's stripes on my shoulder, I felt like a full partner.

They had more questions: Which came out closer to Jessica's room, the elevator or the stairs? Was the emergency stairway any different in design? Were the doors to the patient rooms bulletproof? Did they open in or out? Would the glass panels shatter if fired upon? I answered what I could.

I made ten more maps, all of Saddam Hospital's second floor. I began with the general layout, then zoomed in to the cardiac unit, then to Jessica's room. For comparison, the Marines showed me a set that Iman had helped sketch at the other camp—they were amazing! (In a summary of our work, an officer with the Second Marine Expeditionary Brigade wrote that Iman's maps "were so detailed and organized that they looked much like blueprints of the facility.") My wife had a sharp eye and a strong heart. She might have disagreed with the way I went about things, but she gave her all to help.

Three of Iman's sketches were especially interesting. Years earlier, a top hospital official put through a special construction project: a set of tunnels from the hospital to Guitar Street, so that his mistresses could come to him unseen. The entire staff knew about these secret passageways, which led to the basement as well as to the main floor. Doctors used them to pilfer supplies, nurses to avoid forced overtime. Iman and I had stolen underground kisses there during our courtship. Now the tunnels might serve an entirely different purpose.

The next two days brought more of the same: hours of questions, checks and double checks and triple checks, from early morning to late at night. On Tuesday, April 1, the Iraqis shelled to within fifty yards of the school. The battle for Nasiriya was not yet over. The artillery woke me up, but it no longer startled me. It was like an alarm clock, something half expected.

I got impatient. From the start I'd told Corey and Jim that my information about Jessica was good for only twenty-four hours. Now it had been three days since I had seen her. The fedayeen might have moved her to Baghdad. The doctors might have gone forward with their surgery on her leg. Everything could have changed.

I kept asking Jim, "When are you going to get her?"

And he would say, "I like your urgency—you'd make a good officer. But they have to get all the right people in place." In addition to the Marines, there were Navy Seals, Army Rangers, and Air Force pilots to coordinate, some coming from far away. The Pentagon remembered the disaster in Iran in 1980, when eight soldiers died and no hostages were rescued. Another failure was unacceptable. More than Jessica's life was at stake.

By Tuesday afternoon, I sensed that we were getting closer. Jim asked me about the best time to go in—day or night? I thought early morning would be perfect, when the Baathists were asleep and the fedayeen on duty would be tired.

Then he worked with me to make two more maps of installations to the north and west, where the Iraqi army was guarding the route to Baghdad. The Americans would stage a diversion there before the hospital raid.

Then it was my turn to ask for something. I wanted to go with the commandos. I knew my way around that hospital; I could help them more than any map. I told the master sergeant, "We've lived through some tough moments together, and we've become friends. So let's just keep going together."

They called headquarters in Qatar. Jim came back to me shaking his head: "We can't risk losing you." I would have to wait and watch, the hardest job of all.

When I left him for my bed, shortly after midnight, Jim told me, "It could be tonight or tomorrow. We're just waiting for the okay." The schoolhouse was jammed with young Marines I had not seen before, thirty or more of them, cleaning their weapons in the corridors. *The commandos,* I thought. I took my pill and fell into bed, still in my fatigues.

"Mohammed! Wake up!" A woman's voice, in Arabic— it sounded like al-Hbeiba rousting me for school. I looked at my watch: two-thirty. I tried to focus on the person leaning over me. Ah, it was Lara.

I said, "Go on, I'll meet you." I rolled over.

"Mohammed! We need your help!" She hooked my arm and pulled, not so gently as my mother used to. I blearily put on my sneakers, washed my face, and followed her to the gym.

The corridors were empty again, but every table in the control center was busy. Jim looked up from his desk and gave me the thumbs-up. "Zero hour," he said.

Lara sat me next to her and handed me a headset. We'd be in audio contact with the commandos, she explained. If she needed to ask me something, I should respond by writing yes or no. They didn't want me to speak. A voice in Arabic might confuse things, she said.

For three hours I tuned in to the crackling voices.

There was little I understood, aside from: "Over. Do you read me? Over." It was still dark when I went to the roof for a look. I pointed my binoculars northwest and saw fireworks, from the air and the ground. *The diversion*, I thought. I felt like a coach whose team was playing its biggest game. Now it was up to the players.

I rejoined Lara at her table. She hunched over, listening intently. I could feel a mounting energy in the gym—and then everyone talked at once. Jim looked at me in excitement. *What was going on?* One Marine took off his headset, slammed his table, and said, "God bless you! God bless the U.S.A.!"

People cheered and rushed at me from all angles. They hugged and kissed me, slapped me on the back. They threw me into the air and shouted my name. The operation was a success, they told me. Jessica was in American hands. They said they would bring me to meet her—a meeting postponed when they learned the extent of her injuries.

Lara hadn't needed to ask any questions, she told me, because the maps were perfect. Everything had gone according to plan. The rescue itself had taken six minutes. Though the raiders had been fired upon coming and leaving, they encountered no resistance inside. The Baathists and fedayeen were missing in action. None of us could have known it at the time, but they'd fled the scene the day before.

My first thought was: *It's over.*

And then: *I'll be able to go home now.* The regime was on its knees. Soon life would return to normal—only better, because we wouldn't have Saddam.

It was six in the morning. The Marines relaxed, shedding belts and shoes. I told Lara I was ready to go back to bed. No, she said, I should wait: "A very dear guest is coming to visit you."

"Let me go to sleep," I said. "When he comes, wake me up."

The next thing I knew, my wife was sitting next to me on the bed. Abir jumped giggling into my arms, and then I was *really* awake. Iman said, "Thank God you're here safe—but what happened to you?" She looked anxiously at my eye.

I said, "Everything is fine. I survived." Then I asked, "Did my family make it?"

Iman shook her head. There was no news.

I was distraught. It was four days since I'd said good-bye to Hassan near Baba's house. If a fedayeen or Saddamir had discovered his pass for the Marine camp, it would not matter that Hassan was a policeman. He would die a traitor's death. Baba and Ahmed and the rest might easily be next.

I sank into a dark, dark mood. When word of my trouble got around, a platoon was sent to check Baba's house. Sniper activity was so bad in town that it took them more than twenty-four hours to get there. They returned with a glum report: The house was vacant and ransacked. There was no sign of life.

This was something I had not calculated on. I had focused on Iman and Abir, the fedayeen's obvious targets. Once they were safe, I thought that my father and brothers would have no problem. They had Hassan's police ID, and the pass. The Marines were expecting them. The plan seemed foolproof.

Now I brooded over all I had done wrong. I felt no regret about helping Jessica; I had gone to her with a clear head, because it was the right thing to do. But I should have made sure to save my family first. I should have brought them to the camp myself. When the people you love are at stake, you cannot rely on anyone else.

To rescue Jessica was a great victory, and I would always be proud of my part in it. But the cost—oh, yes, the cost was great to me as well.

On April 4, two days after the rescue, we left the school-house for Jalibah Airfield, in the middle of the Iraqi desert. It was a short trip by helicopter, but the skies were not safe, and Iman held on to her seat with white knuckles. Abir was frightened by the noise and cried all the way. She cheered up when we got to Jalibah and were showered with gifts. There was a huge green pullover for me, a Marine T-shirt for Abir, and an MTV T-shirt for Iman.

After one night in an RV, we flew the next day to the British airfield in Umm Qasr and another warm welcome. As we whiled away the next two days, I amused myself by training the troops in kung fu, until the doctor made me stop to protect my eye. The pain was subsiding, little by lit-tle, but I still had no vision there.

On April 7, thanks mainly to the efforts of Lt. Col. Bill Perez and Capt. Jason Berg, we were granted refuge to leave Iraq. Our next stop was Kuwait City, where the U.S. ambas-sador, Richard Jones, was waiting at the embassy door to bid me "Welcome to America in Kuwait." I'd hoped to wear something nice for the occasion, but my green suit was a mess. I settled for camouflage pants and my pullover, with a floppy camouflage hat I'd grown fond of.

We stayed in a large apartment in the embassy residence. There was a gym, a swimming pool, even room service. It was a wonderful place, but I cannot say that I enjoyed myself. I worried constantly about my family, and I dreamed of blood and death. I saw the Head Cutter, and the mur-dered businessman calling my name, over and over . . .

Once I dreamed of al-Hbeiba. She smiled sadly and said, *You left Iraq without saying good-bye to me.*

The morning after our arrival, I met with an embassy lawyer to discuss my family's future. I knew that U.S. infantry had taken Saddam International Airport and encircled Baghdad—the war would soon be done. We would stay in Kuwait for two months, I thought, and then go back to pick up our lives. I had property and a business to tend to. My house would need work, but furniture could be replaced. Nasiriya had been our home for generations. Where else would I go?

The lawyer saw it differently: "Things are not going to settle down in two months. It could easily be a year. Right now it looks like it could easily be five years."

As we talked, I saw that I had been fooling myself. The fedayeen would not evaporate with the fall of Saddam. They had long memories; if the wrong people crossed my path, they would kill me in a blink. There were also the fundamentalist Shiites, who had never been kind to collaborators. My story was out on the news wires. I withheld my last name from the press, but those I had angered knew me. They would be ready if I returned.

I was a child of Iraq, and that could never be erased. On the other hand, I had always been out of step there, ever since first grade. Now I had taken one giant step too far. I wished the best for my country—for an end to the bloodshed, above all. But I understood that my old life was over.

The American government offered my family "humanitarian parole," and soon we would be granted asylum. According to Sean Murphy, the U.S. consul in Kuwait, we could live almost anywhere in the world. I had relatives in Denmark, Sweden, and Australia. I had my brother in Jordan, where my family owned a restaurant.

Sean said, "We'll settle you wherever you want. Take your time—you don't have to decide right now."

That had been my intention, to tell Sean and his staff

that I needed time to think. But in that moment I was absolutely sure of where I wanted to go. The fedayeen could reach across the border to Jordan, where the Americans' war had been unpopular. Those other countries were words on a map, pictures in magazines. They meant nothing to me.

For two strange, long weeks I had lived and worked with Americans. I liked everything about them. They were honest and well organized, and most of all they were human. They had treated me warmly. They had made me feel like I was one of them.

I had no job waiting for me in the United States, no place to stay, not a single person I knew. I spoke the language poorly. I knew that I would struggle there.

But there was no place I would rather be—to raise Abir in safety, to live without fear. I told the consul, "I started this with the Americans, and I want to finish it with the Americans."

Everyone seemed to think that was a perfect answer.

Early the next morning, Iman, Abir, and I said our good-byes. With little more than the clothes on our backs, we boarded a military jet for a place we had seen only on television and in the movies. It would be a long journey, and our future was uncertain. But I was comfortable with that.

In a funny way, I felt that we were going home.

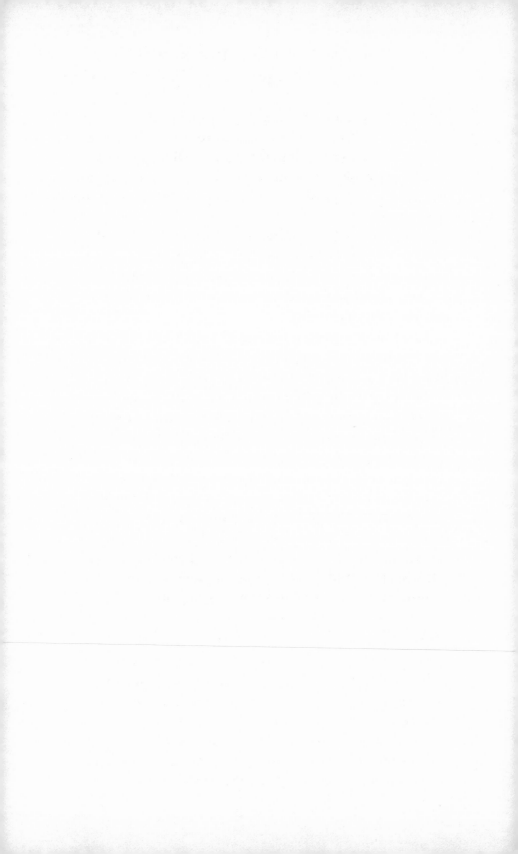

EPILOGUE

SINCE COMING TO AMERICA in April, I have been met by kindness and generosity wherever I have gone. It began with the women of the International Relief Committee, who bought me two suits after we touched down in Boston. Then I met Bob Livingston in Washington, D.C. He reminded me of Baba—a man of strength, but also compassion. When he offered me a job with the Livingston Group, an international lobbying firm, my new life became something rooted.

Through Bob I met a Lebanese-American lawyer who graciously opened his home to our family. The lawyer put me in touch with Barrett Katz, chairman of the ophthalmology department at the George Washington University Medical Center. Barrett referred me to Mark Johnson, one of the top retinal surgeons in the country. Along with Suburban Hospital in Bethesda, Mark offered his services free of charge.

The shrapnel was only two millimeters long, but it made its presence felt. It had entered through the side of my left eyelid, bounced though the lens, torn the retina, and lodged in my iris. On May 5, in a three-hour surgery, Mark removed the shrapnel and reattached the retina with a laser spot weld. The operation was successful, but there were no guarantees. The retina—"the film in the camera," as Mark put it—was permanently damaged. I might get some small amount of my sight back, or nothing at all.

It was a tough adjustment at first. I had little depth perception and kept walking into people. Once I went into a restaurant and knocked a drink from a woman's hand, dousing the both of us.

On a sunny day in mid-June, I was working in a hotel suite, facing a large window, when I *felt* my pupil contract. I closed my right eye—there was light out there! When I passed my hand in front of my face, I could see a faint shadow, then light again. It was an amazing sensation.

I've noted little progress since. With time, Mark says, I may be able to see motion and perhaps some vague shapes, but no more. In any case, I am learning to compensate. I can pour a cup of coffee without scalding myself, and soon I will be able to drive. I am told that 10 percent of the American population sees with only one eye. Now I am one of them.

After a month in this country with no word, I had nearly given up hope for my family in Iraq. The guilt pressed on me every day; I would take it with me to the grave, I thought. When the call came into the Livingston Group office, I would not believe it until I heard my brother Ahmed's voice.

"We're okay!" he said. "We are all with the Americans—there is no worry!"

It is hard to describe my feelings. I shouted my brother's name, just to hear him respond—*it was him!* Just like that, seven people came back to life.

No, *eight* people came back, because I was reborn.

When I stopped shouting, I listened to Ahmed's story. They had hidden for a while at the farm, he said, and later at their vacant "safe house" in town. When they sensed that their neighbors were on to them, they left for a Marine camp west of the city. That move may have saved their lives. A few days later, regime loyalists came and burned the house down.

My joy was short-lived. As I kept in touch with Ahmed over the next several weeks, whenever he got access to a phone, the news got darker. My family was sleeping in a simple tent, as hot as a furnace. My niece got bitten by a scorpion. Baba's heart was flagging, Ahmed said. His legs were swollen, his breathing labored. Hamida was doing the best she could with him, and eventually the family was moved into air conditioning. But Baba still struggled with his heart and diabetes, and lost some of his sight. He needed top-flight hospital care; he needed to come to the West. It was not a simple thing to make happen.

As I write this, Bob Livingston and other friends are working to gain asylum for my family. I look forward to the time we are together again—perhaps, someday, as American citizens.

My father is seventy-six years old, a difficult age to see your life turned inside out. My hope is to regain some peace for him, to make up for the trouble I've caused. Baba and I have yet to speak since I left Iraq, and that day cannot come too soon. Until then, I will hold on to a message sent through Ahmed: *He is proud of you, Mohammed.*

Coming from a sheikh, and the son of a sheikh, that meant something.

Iman misses her family and struggles with all the changes in our lives. But she has a plan to study English and go back to nursing. She is accustomed to challenges—she married me, after all. She has great willpower, and I know she will succeed.

Abir still gets tired when she runs or uses the stairs. She has nightmares about people burning; she'll shrink back when someone lights a match. But I also see my daughter healing, laughing more day by day. She sponges up her new language from playmates and cartoons. This fall she starts kindergarten. She will be a true Iraqi-American.

I read everything I can about Jessica's progress. When the hospital released her to her home in West Virginia, I felt like we had won all over again. More than anything, I wish her a full and happy life.

And if I were to meet her, what would I say? I would tell her that I share America's pride in her courage. I would tell her that we are two of the fortunate people alive today, simply because we are *alive*. I know this whenever I look at the American flag on my bedroom wall, given to me by the Marines, and remember that it might have been my shroud.

Finally, I would tell Jessica this: I regret nothing. I would risk it all again to help her, without thinking twice. And I would thank her, because I have gained more than I lost. I have gotten back what Saddam tried to take from us all those years.

I have my humanity, and that is enough.